COMBINATORIAL ORGANIC CHEMISTRY: AN EDUCATIONAL APPROACH

COMBINATORIAL ORGANIC CHEMISTRY: AN EDUCATIONAL APPROACH

COMBINATORIAL ORGANIC CHEMISTRY: AN EDUCATIONAL APPROACH

SHERIF EL-BASIL

Nova Science Publishers, Inc.
Huntington, New York

Editorial Production: Susan Boriotti
Office Manager: Annette Hellinger
Graphics: Frank Grucci and Jennifer Lucas
Information Editor: Tatiana Shohov
Book Production: Donna Dennis, Patrick Davin, Christine Mathosian, Tammy Sauter and Lynette Van Helden
Circulation: Maryanne Schmidt
Marketing/Sales: Cathy DeGregory

Library of Congress Cataloging-in-Publication Data

El-Basil, Sherif.
Combinatorial Organic Chemistry:An Educational Approach / Sherif El-Basil
p. cm.
Includes bibliographical references and index.
ISBN 1-56072-744-6
1. Combinatorial chemistry. 2. Stereoisomers. I. Title..
RS419.E4 1999 99-049902
615'.19--dc21 CIP

Nova Science Publishers, Inc.
227 Main Street, Suite 100
Huntington, New York 11743
Tele. 631-462-6682 Fax 631-462-4666
e-mail: Novascience@earthlink.net
e-mail: Novascil@aol.com
Web Site: http://www.nexusworld.com/nova

Printed in the United States of America

Contents

Preface

This is an educational book which considers several isomer-enumeration methods in organic chemistry .Three main topics being exemplified here , viz., theorem of **pólya** , coset representation theory and marks of a group(method of **Fujita**) and wreath and generalized wreath product groups (method of **Balasubramanian**) .The book extents an invitation to the class of chemists who are repelled by formula language but who may be interested to learn how to apply combinatorial theory to stereochemistry . Consequently rigorous proofs have been replaced by an example-oriented , intuitive approach which exposes many problems of **real** chemical importance . The stereochemistry of the **lactic and tartaric acids** is dealt with in **examples 5 and 35** respectively. Stereoisomerism of **cyclopropanes and cylcobutanes** is treated in **examples 6,7,10** while **open-chain geometrical isomers** are studied in **example 8**.The mathematical origin of the **axial and equatorial bonds** in cyclohexane is considered in **example** 9.The stereochemistry /combinatorial theory of the molecules of **allene , cubane and trishomocubane** is illustrated in **examples 11, 12 and 37** respectively .The inorganic **octahedral** isomers are studied in **example 13** . Several classes of **heterocyclic compounds** are combinatorially generated in **examples 14-20 , 33, 34 and 36.**

The (sometimes) "offensive" algebra needed to understand the **enumeration "journey" of Fujita** (c.f., **FIG 36)** is modeled using a set of graphs such that any reader who can only rotate and/or reflect objects will certainly be able to communicate abstract terms of the theory such as coset representation , group-subduction , mark table and many others. **FIGS 1-3** model identity representations , **G(/G)**'s , for fifteen chemically important pont-groups. **FIGS 46-75** model other concepts of the theory of cosets in a chemically appealing form , For example **FIG 67** is a pictorial illustration of the subduction of a coset representation of a group by one of its subgroups. When abstract symbols turn into observables (such as objects , graphs etc.) many of the opaque features of a theory become apparent. Similarly **Table 35** models the **mark table** of the **D_3** point group where the reader , perhaps for the first time recognizes numbers (in the conventional form of such tables) as colored graphs and whence realizes the physical meaning of these tables.Indeed modeling the algebra of coset representation theory makes many of the algebraic results easy to derive. This is done in **Section 4.4** of the book.

Analogously we used simple graphs, called caterpillar trees to model the (involved) combinatorial realtions in wreath (and generalized wreath) product groups necessary to understand the study of stereoisomerism using the theory of Balasubramanian .**Examples 53 and 54** illustrate these concepts in simple terms.

The topic of nmr (both pmr and ^{13}C nmr) is treated using the three methods , viz., Pólya , Fujita and Balasubramanian .These approaches are exemplified by the benzenoid hydrocarbons :naphthalene , anthracene , pyrene and coronene (c.f. **examples** 21-30, 38 and 55).

All the above and many other examples are casted in details , step-by-step so that any reader will definitly gasp all aspects of the method to the level of **how-to-do** other problems. Hopefully readers of this primer will be motivated to read more rigorously written formal treatments, and in particular the organic chemist will eventually find his way to the **heavens of Fujita!** (c.f. **FIG** 35).

Cairo , Egypt

April 1999

Sherif El-Basil

Chapter 1

Review of Some Elementary Concepts of Symmetry

1.1 Symmetry operations and symmetry elements.

A given three-dimensional object, which may represent a chemical structure has certain number of **symmetry operations**. These are defined to involve movement of the configuration such that after the movement has been carried out every point of that configuration is coincident with an equivalent point, ie., the resulting configuration cannot be distinguished from the original one. Imagine, e.g., the molecular model of (hypothetical) cyclobutadiene which is simply (nearly) a square. Now consider an axis which is perpendicular to the plane of the model and rotate the model around this axis by an angle = 90°. We observe that the rotated model is identical with the original one. This operation of rotation is called a symmetry operation. We observe that symmetry operations leave points (centers), lines (axes) or planes invariant (i.e. fixed). These latter objects which remain fixed after a symmetry operation are called **symmetry elements**. In our previous example, the axis through which the model has been rotated is certaintly one of the elements of symmetry of this particular configuration. As a matter of convention symmetry operations are expressed using italicized (Greek) symbols while symmetry elements are written using usual capital letters. These concepts are explained in details in several books[1-5] and we see no point in duplicating information here. Throughout this book we adopt the following (Greek) symbols to describe symmetry operations as follows:

$I = E$: identity operation. E.g., rotation around an axis by 360° or reflecting an object twice.

C_n : rotation by an angle $\theta = 360/n = 2\pi / n$.

σ_h : reflection in a horizontal plane.

σ_v : reflection in a vertical plane.

σ_d : reflection along a diagonal plane.

i : inversion through a center.

S_n : rotoreflection by an angle $\theta = 2\pi / n$ i.e. rotation followed by reflection, i.e., $S_n = \sigma_h C_n$ where σ_h must be perpendicular to C_n.

The reader may wish, at this point, to refer to FIG 1-3 where these definitions are exemplified (See Section 1.2).

1.2 Group-theoretical structure of symmetry operations

The set of elements which form all the symmetry operations of a given configuration possesses the mathematical properties of a (point) group, **G**, i.e.,

a) The product of two elements of **G** is also an element of **G** ;

b) Let a,b,c,.. be elements which belong to **G**, then multiplication of any three such elements is *associative*, i.e., (ab)b = a(bc) ;

c) There is an identity element I (or E) in **G** which has the property that gI = Ig =g, where g is an element of **G**.

d) For every element g in **G** there is an *inverse* element g^{-1} such that $gg^{-1} = I$

The above properties can be tested by examination of the *multiplication table* of a given point group. As an illustration we show in Table 1 the multiplication table of the $\mathbf{D_4}$ group [C.f. FIG 1].

Table 1

Multiplication table of $\mathbf{D_4}$ group.

2nd Operation \ 1st operation	I	C_4	C_4^2	C_4^3	C_2	C_2'	$C_2(d)$	$C_2'(d)$
I	I	C_4	C_4^2	C_4^3	C_2	C_2'	$C_2(d)$	$C_2'(d)$
C_4	C_4	C_4^2	C_4^3	I	$C_2(d)$	$C_2'(d)$	C_2'	C_2
C_4^2	C_4^2	C_4^3	I	C_4	C_2'	C_2	$C_2'(d)$	$C_2(d)$
C_4^3	C_4^3	I	C_4	C_4^2	$C_2'(d)$	$C_2(d)$	C_2	C_2'
C_2	C_2	$C_2'(d)$	C_2'	$C_2(d)$	I	C_4^2	C_4^3	C_4
C_2'	C_2'	$C_2(d)$	C_2	$C_2'(d)$	C_4^2	I	C_4	C_4^3
$C_2(d)$	$C_2(d)$	C_2	$C_2'(d)$	C_2'	C_4	C_4^3	I	C_4^2
$C_2'(d)$	$C_2'(d)$	C_2'	$C_2(d)$	C_2	C_4^3	C_4	C_4^2	I

Suppose, e.g., we wish to find the product $C_2C_{2(d)}$. *In this case* $C_{(d)}$ is the 1st operation while C_2 is the 2nd operation. We can perform these two operations successively on the model shown in FIG 1 or a simpler model for this purpose. Such a simplified model would be a square (with four vertices and four edges). The first operation is reflection of the square along one of its diagonals, viz., $C_2(d)$ while the second operation involves rotation of the resulting configuration along C_2 axis as pictorially illustrated in eqn. (1):

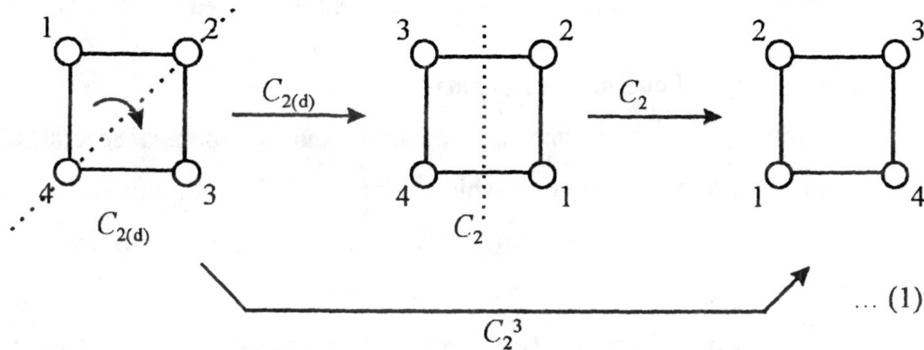

... (1)

The two operations are equivalent to rotation along an axis which is perpendicular to the plane of the model by an angle of 270° in a clockwise fashion. Whence we write

$$C_2\, C_{2(d)} = C_4^3 \qquad \dots (2)$$

This result can be obtained fromTable 1.1. In general (by convention) multiplication table are designed so that a given product is the result of a column element times a row element. The column element represents the 2nd operation while the row element represents the 1st operation. The *order* of operations is quite important here because in general this type of multiplication is not commutative. (i.e., ab = ba). Multiplication tables of point groups may be found scattered in texts dealing with symmetry but quite often the "*desired*" table may not be found and because such tables are crucial to the enumeration techniques, as it turns out, it is strongly adviced that a reader must be trained of constructing equations such as eqn.(1). For table 1, one needs 8 x 8 = 64 such equations. However the first row is always trivial because it involves identity multiplications.

1.3 Subgroups of a given group

The set of elements { $g_1, g_2, \ldots, g_{|G|}$ } = **G** where |**G**| is the order of group **G**, i.e. the number of elements in it, can often be decomposed into subsets, each of which satisfies the axioms (properties) of a group mentioned in Section 1.2. Each such subset is called a *subgroup* of the original group. In FIG 1-3 many point groups are modeled along with their subgroups. It turns out that finding all subgroups of a given group is the key step in combinatorial enumeration of various types of isomers.

1.4 Some point groups of chemical relevance

FIG 1 pictorially illustrates symmetry operations and subgroups of several of the C_n groups for n = 2,3 and 4. Molecules which belong to this point group possess a C_n axis in addition to the identity. A molecule which belongs to $\mathbf{C_n}$ possesses n symmetry operations. C_n axes are perpendicular to molecular plane and rotations are assumed to take place in a clockwise fashion. The addition of n two-fold axes at right angles to the C_n axis generates the point group $\mathbf{D_n}$. FIG 1 illustrates $\mathbf{D_2}$, $\mathbf{D_3}$ and $\mathbf{D_4}$. Because the point group $\mathbf{C_3}$ exists as a subgroup of $\mathbf{D_4}$, it is also modeled in FIG 1.

Figure 1 (see page 153)

The addition of one vertical plane of symmetry, σ_v , to the C_n axis of a $\mathbf{C_n}$ requires the addition of (n-1) σ_v planes and the resulting point group is denoted as $\mathbf{C_{nv}}$. In FIG 2 we consider this point group for n=2 and 3. The addition of a horizontal plane to a $\mathbf{C_n}$ leads to the point group $\mathbf{C_{nh}}$. In FIG 2 the $\mathbf{C_{3h}}$ is illustrated. When n planes of symmetry are added to a $\mathbf{D_n}$, the point group $\mathbf{D_{nd}}$ results. Finally the addition of a horizontal plane to the elements of $\mathbf{D_n}$ generates the point group $\mathbf{D_{nh}}$. FIG 2 exemplifies both $\mathbf{D_{3h}}$, $\mathbf{D_{2d}}$ and $\mathbf{D_{3d}}$. Because $\mathbf{S_4}$ appears as a subgroup for $\mathbf{D_{2d}}$, it is also shown.

Figure 2 (see page 154-155)

FIG 3 considers an examples of the cubic group, **T** modeled by the molecular graph of adamantane. This important point group contains $\mathbf{C}_2$, $\mathbf{C}_3$ and $\mathbf{D}_2$ as subgroups.

Figure 3 (see page 156)

These point groups will be used throughout Chapter 3 to model symmetry and combinatorial enumeration in chemistry.

Bibliography

Books 1 and 2 are educational, 3-5 are more mathematically oriented:

1. G. Davidson, Group theory for chemists, Macmillan educational LTD, London (1991).
2. A. Vincent, Molecular symmetry and group theory, John Wiley & Sons, New York (1996).
3. R.L. Flurry, Jr, Symmetry groups, Prentice-Hall, Inc., Englewood Cliffs, New Jersey (1980).
4. H.H. Jaffé and M. Orchin, Symmetry in chemistry, John Wiley & Sons, New York (1965).
5. B.S. Tsukerblat, Group theory in chemistry and spectroscopy, Academic press, London (1994).

Chapter 2

Chemical Enumeration Using Theorem of Pólya[1,2]

2.1 Permutation Representation of Symmetry Operations.

Our objective now is to show how a given set of symmetry operations leads to a set of permutations of the numbers 1,2,…,n where n is the number of positions to be moved or stay fixed (invariant) under a given symmetry operation. Usually it is convenient to use the molecular graph which represents a chemical structure. For example if one is interested in the carbon skeleton of (hypothetical) cyclobutadiene one uses the symmetry of a square. This particular model has the 8 symmetry operations characteristic of the $\mathbf{D_4}$ point group [c.f. FIG 1]. When the four vertices of the square are labeled as 1,2,3,4 , (n=4 in this case), and the set of (eight) symmetry operations are applied to it one generates the permutations shown in FIG 4.

Figure 4 (see page 157)

In general when we have a cycle (such as the square) the labeling of its vertices is done in a clockwise way. Also rotations about axes which are perpendicular to paper plane are performed in a clockwise way. The set of symmetry operations shown in FIG 4 is divided into rotations and reflections. The notation (1432), e.g., associated with the operation of C_4 means that 1 (i.e. vertex labeled 1) is sent to 4, 4 is sent to 3, 3 is sent to 2 and finally 2 is sent back to 1, thus completing the 4-cycle. An equivalent notation is

$$\begin{pmatrix} 1 & 2 & 3 & 4 \\ 4 & 1 & 2 & 3 \end{pmatrix} \quad \dots (1)$$

The permutation (1432) is a short thand notation of the 4-cycle shown below:

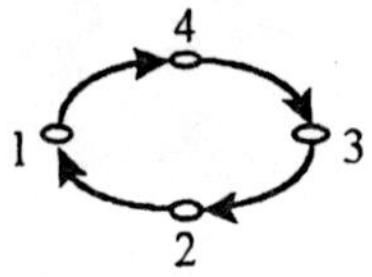

4
1 3
2
… (2)

Analogously the permutation expressed as (13) (24) represents the 2-cycles, viz.,

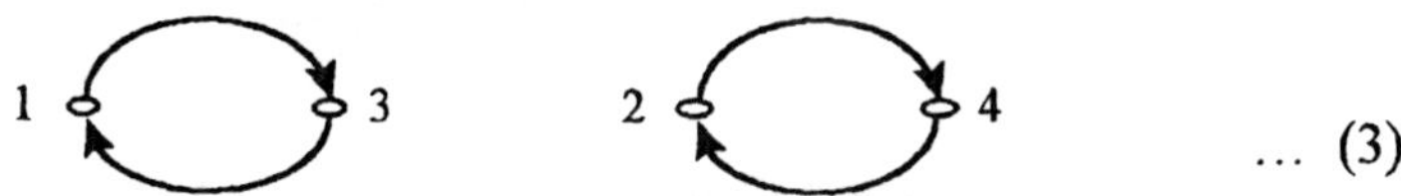

... (3)

To further explain the meaning of the notation we consider the permutation (1) (3) (24) resulting from diagonal reflection of the square [C.f. FIG 4]. This notation is equivalent to two 1-cycles and one 2-cycles, viz.,

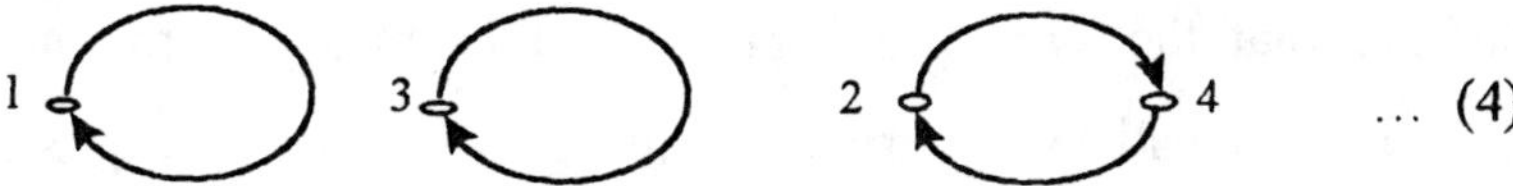

... (4)

The above notations are usually expressed in terms of a dummy variable given by eqn. (5), viz.,

$$S_i^j \qquad \ldots (5)$$

where the subscript i = the length of the cycle. For example a 1-cycle corresponds to i=1, a 2-cycle corresponds to i = 2 and so on. The superscript j is the number of cycles which possess identical lengthes. As an illustration we consider the following cases:

$$(1\;4\;3\;2) \quad = \quad S_4^1 \;=\; S_4\,, \qquad \ldots (6)$$

$$(1\;3)\;(2\;4) \quad = \quad S_4^2 \qquad \ldots (7)$$

$$(1)\;(3)\;(2\;4) \quad = \quad S_1^2\;S_2 \qquad \ldots (8)$$

For the last case, eqn.(8), one must use two S_i^j symbols: one to describe the two one-cycles and the other to describe the one two-cycle. (remember that a one-cycle (= 1-cycle) means a cycle of length = 1 as cycles 1 and 3 in eqn. (4), a two-cycle (= 2-cycle) is a cycle of length = 2 and so on).

2.2 The Cycle Index

When the S_i^j symboles are summed and divided by the number of symmetry operations one obtains an expression for the cycle index. There are two types of cycle-indices:

a) A cycle index involving *rotation only*. This is referred to as a 2-dimensional cycle index and is given the symbol Z_2 .

b) When the expression involves *both rotations and reflections*, a 3-dimensional cycle index, Z_3

For example, for the square shown in FIG 4 we can write the following expressions for Z_2 and Z_3:

$$Z_2 = 4^{-1} \; [\; S_1^4 + S_2^2 + 2\, S_4] \qquad \ldots (9)$$

$$Z_3 = 8^{-1} \; [\; S_1^4 \; + 3\; S_2^2 \; + 2\; S_4^1 \; + 2\; S_1^2\, S_2^1 \;] \qquad \ldots (10)$$

We observe that the sum of the product of i times j is the number of positions operated upon by all symmetry operations (=n). For example eqn. (10) has four distinct terms each of which leads to n = 4:

$$S_1^4 \;\rightarrow 1 \times 4 = 4 \;\; ; \;\; S_2^2 \rightarrow 2 \times 2 = 4 \;\; ; \;\; S_4^1 \;= 4 \times 1 = 4\,;$$

$$S_1^2\, S_2^1 \rightarrow \; 1 \times 2 \; + \; 2 \times 1 \; = \; 4 \qquad \ldots (11)$$

References 1 and 2 consider cycle indices.

FIG 5 illustrates symmetry operations of the molecular graph of naphthalene and the cycle-notations.

Figure 5 (see page 158)

FIG 6 does the same for the molecular graph of pyrene.

Figure 6 (see page 158)

2.3. Pólya's Method of Enumeration: Equation of Burnside, The Generating Function[2]

We learn how to use the theorem of Pólya via the following (almost trivial) example : Consider the skeleton shown below:

$$\ldots (12)$$

where X = {C, N}. How many structures are consistent with the above general formula? The question may be casted in a slightly different way, viz., how many *colorings* result when the vertices of a square are colored in two ways, say, black and white? Indeed the language of "coloring" is the "mathematical analog" of the chemical term: *substitution*. Here there are two colors, viz.,

$$X_1 = C \quad ; \quad X_2 = N \qquad \ldots (13)$$

At this point, it is convenient to introduce an important variable, m = the number of colors = 2 in this particular case. To find the required answer we carry out two steps:

a) Find the cycle index of the *parent*, i.e., unsubstituted (= uncolored) graph. This is simply a square (= 4 vertices linked by four edges in a cycle). This has already been dealt with in FIG 4 and the cycle indices Z_2 and Z_3 are given by eqns. (9) and (10) respectively. In general for enumeration problems we use Z_3

b) We then use a simplified version of the so-called *theorem of Burnside*[1,2] given by

$$\text{Number of } \textit{distinct} \text{ structures} = \frac{1}{|G|}\sum m^j \qquad \ldots (14)$$

Where j is the number of cycles [C.f. eqn. (5)] and the summation is taken over all terms in the expression of the cycle index (i.e. over all symmetry operations). For the case at hand we re-write Z_3 of the sauqre, viz.,

$$Z_3\,(\text{square}) = 8^{-1}\,[\,S_1^4 + 3\,S_2^2 + 2\,S_4^1 + 2\,S_1^2\,S_2^1\,] \qquad \ldots (10)$$

Now, eqn. (14) modifies the terms in eqn. (10) as follows:

$$S_1^4 \rightarrow m^4 = 2^4$$

$$3\ S_2^2 \rightarrow 3.m^2 = 3.2^2$$

$$2\ S_4^1 \rightarrow 2.m^1 = 2.2$$

$$2\ S_1^2\ S_2^1 \rightarrow 2.m^2.m^1 = 2.2^3$$

... (15)

Then the number of distinct structures is given by :

$$8^{-1}\ [\ 2^4 + 3.2^2 + 2.2 + 2.2^3\ ] = 6$$

The six structures are identified as: one structure containing only C, abbreviated as C^4, one structure which has one C atom and three N atoms, abbreviated as CN^3, two structures each of which has two N and two C atoms, one structure containing three C atoms and one N atom and finally one structure which has four N atoms. The situation may be denoted as :

$$C^4 + CN^3 + 2\ C^2N^2 + C^3N + N^4 \quad \text{... (16)}$$

(See also, FIG 7, below)

Actually the expression reached in eqn. (16) is called the *generating function*, GF, which can be obtained when we apply the following, so-called, Pólya's substitution to each term in Z_3, viz.

$$S_i^j \rightarrow (\ X_1^i + X_2^i + \ldots + X_m^i\ )^j \quad \text{... (17)}$$

where i = cycle-length (C.f. eqn. (5)] , j = number of cycles and X_1 is called the "weight" of the first color and so on. For the simple case at hand we have:

$$S_i^j \rightarrow (\ X_1^i + X_2^i\ )^j \quad \text{... (18)}$$

and taking $X_1 = C$ and $X_2 = N$ the terms in eqn. (10) generate the following polynomials:

$$S_1^4 \rightarrow (C+N)^4 \quad ; \quad S_2^4 \rightarrow (C^2+N^2)^2\ ;$$

$$S_4 \rightarrow (C^4+N^4) \quad ; \quad S_1^2\ S_2 \rightarrow (C+N)^2\ (C^2+N^2)$$

... (19)

Where our generating function is given by :

$$GF = 8^{-1} [(C+N)^4 + 3 (C^2 + N^2)^2 + 2 (C^4 + N^4) + 2(C+N)^2 (C^2+N^2)]$$

$$= C^4 + CN^3 + 2 C^2 N^2 + C^3 N + N^4 \qquad \ldots (20)$$

which the same expression reached before by simple inspection, eqn. (16) For "real" problems, however, it is not always possible to find the required structures and one must expand the expression for the generating function.

2.4 Equivalence Classes

A beginning student may erroneously think of the problem of the square just posed above in the following way: Each vertex has two colors available viz., C and N and whence the total number of structures is $2^4 = 16$ (because there are 4 vertices in the square). Actually this type of "reasoning" counts too many structures! In fact many of the resulting structures will be "equivalent" to one another. *In chemistry* two structures are considered equivalent if one can be obtained from the other by a symmetry operation. A (sub)-set of equivalent structures is called an *equivalence class*. An equivalence class then corresponds to one of the *distinct* structures. Whence we expect six equivalence classes for the square problem. These classes are shown in FIG 7.

Figure 7 (see page 159)

2.5. Chemical Applications of Polya's Theorem,

2.5.1 Mathematical origins of stereochemical terms

In this section we consider several examples which illustrate the mathematical origins of enantiomers, diastereomers, axiaL bond, equatorial bond, etc. It is our intention to solve as many examples as possible instead of "locking-up" informations in (usually unsolved) exercises! First we shall begin with simple (almost trivial) examples, then we gradually move on to more realistic stereochemical problems.

Example 1 **The equilateral triangle**

Find the number of structures which possess the following general formula

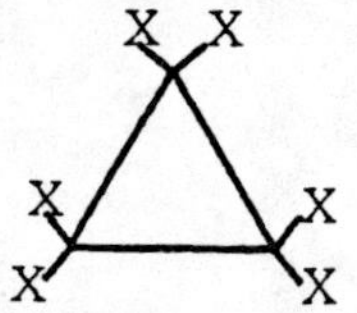

... (21)

where X = { H, Cl }. Draw these structures.

First we recognize that the parent graph is an equilateral triangle i.e. three vertices linked by three edges in the form of a cycle. The point grou is a C_{3v} (C.f. FIG 2). The six symmetry operations, their permutation representations and the cycle index are all portrayed in FIG 8.

Figure 8 (see page 160)

The expression of Z_3 is given by :

$$Z_3 \text{ (triangle)} = 6^{-1} [S_1^3 + 2\, S_3^1 + 3\, S_1^1\, S_2^1] \qquad \text{... (22)}$$

Naturally if the superscript j = 1 one can drop this superscript and whence the second and third terms of eqn. (22) may simply be written as S_3 and $S_1\ S_2$ respectively.

The number of distinct structures (= the number of equivalence classes) is given by eqn. (14) , Burnside's eqn., viz.,

$$6^{-1}\ [\, 3^3 + 2.3^1 + 3.3^2\,] = 10 \qquad \text{... (23)}$$

Observe that m (= number of colors) = 3 (not 2 as might be anticipated as only H and Cl !). The tree colors are shown below:

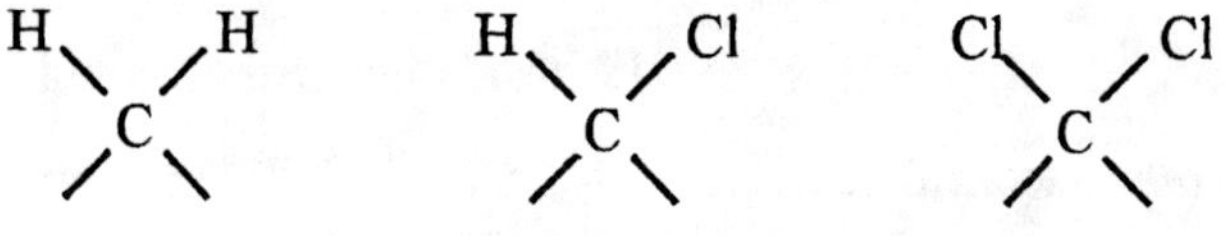

$X_1 = a$ $X_2 = b$ $X_3 = c$ ... (24)

where we have given them the symbols a,b and c. To find these structures we derive the generating function by applying Pólya's substitution, eqn. (17) to the expression of Z_3 , eqn. (22):

$$GF = 6^{-1} [(a+b+c)^3 + 2(a^3+b^3+c^3) + 3(a+b+c)(a^2+b^2+c^2)]$$

$$= a^3+b^3+c^3+ab^2+ac^2+bc^2+a^2b+a^2c+b^2c+abc \quad \dots (25)$$

These structures are drawn in FIG 9 where the expanded form of eqn. (25) contains shorthand notations for the actual structures as shown in the figure. Expressions such as the one given by eqn. (25) are sometimes called *pattern inventory* which contains the generating function of all possible (distinct) structures.

Figure 9 (see page 161)

Example 2 **The Square**

Find the number of structures (neglecting stereochemistry) which bossess the following general formula

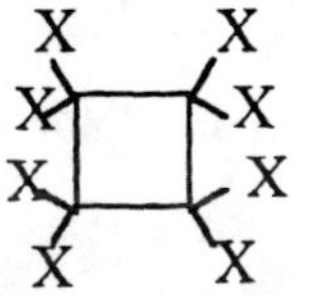

X X X X X X X X ; $X = \{H, Cl\}$ ... (26)

Derive a generating function for these structures and draw some of them.

As in the previous examples m = 3 because there are three types of substituted C atoms as in eqn. (24). The cycle index of the square in three-dimensions has already been derived [c.f. eqn. (10) and FIG 4]. Then, the number of structures* is given by:

$$8^{-1} [3^4 + 3.3^2 + 2.3^3 + 2.3] = 21 \quad \dots (27)$$

The expression for the generating function allows a display of these 21 structures, viz.,

$$GF = 8^{-1} [(a+b+c)^4 + 3(a^2+b^2+c^2) + 2(a+b+c)^2(a^2+b^2+c^2) + 2(a^4+b^4+c^4)]$$

*From now on . number of structures means number of distinct structures

$$= a^4 + b^4 + c^4$$
$$+ 2a^2b^2 + 2a^2c^2 + 2b^2c^2$$
$$+ ab^3 + ac^3 + bc^3 + a^3b + a^3c + b^3c$$
$$+ 2abc^2 + 2ab^2c + 2a^2bc$$
... (28)

where a, b and c are the three possible substitution patterns (= colors) defined in eqn. (24).

The straightforward (but somewhat tedious) algebra required to expand the above (and similar) GF`s is facilitated using multinomial theorem.[1]
It is instructive to draw the last six structures [in eqn. (28)]. These are drawn in FIG 10.

Figure 10 (see page 162)

At this stage it may be convenient to examine the *effect of the third dimension*, namely by using Z_2 instead of Z_3. We recall that Z_2 contains only rotations while Z_3 contains both rotations and reflections. Whence we expect that Z_3 "sees" two structures which are related by a mirror plane as *one structure*. On the other hand Z_2 (which does not include reflections) will count two enantiomers as two distinct structures!. We shall postpone the use of Z_2 in counting the various stereoisomers until later, but for now we illustrate the effect of the third dimension by the following simple calculation.

Example 3 The Effect Of The Third Dimension

We repeat the above calculation [of coloring the vertices of the square by the three colors a, b and c of eqn. (14)] using Z_2, i.e., the cycle index in two dimensions. This is given by eqn. (9) and FIG 4. Using m = 3 leads to the following count:
Number of structures in 2-dimensions

$$= 4^{-1} [3^4 + 3^3 + 2.3] = 24$$ ... (29)

Then the removal of elements of reflections, viz., σ_v, σ_h, $\sigma_{d's}$, leads to three extra structures. These "extra" structures can be guessed by expansion of the relevant generating function. We have :

$$
\begin{aligned}
GF &= 4^{-1} [(a+b+c)^4 + (a^2+b^2+c^2)^2 + 2(a^4+b^4+c^4)] \\
&= a^4 + b^4 + c^4 + a b^3 + a c^3 + b c^3 \\
&\quad + a^3 b + a^3 c + b^3 c \\
&\quad + 2 a^2 b^2 + 2 a^2 c^2 + 2 b^2 c^2 \\
&\quad + 3 a^2 bc + 3 a b^2 c + 3 a b c^2
\end{aligned}
\qquad \ldots (30)
$$

Then the "extra" 3 structures are of the type $a^2 bc$. The absence of reflection leads to counting two equivalent structures as two distinct ones! In our case structures of the type $a^2 bc$ in which the two a`s are adjacent are counted twice as shown in FIG 11.

Figure 11 (see page 162)

But one may wonder: Why in the absence of reflection structures of the type $a^2 bc$ in which the a`s occupy diagonal positions are not doubled ? To answer this question let us reflect one such structure as shown in FIG 12. We will then see that the reflected structure can also be obtained by rotation of the original figure by 270°,i.e. by C_4^3 which is already present in Z_2 and whence the two (equivalent) structures are counted (by Z_2) as one.

Figure 12 (see page 162)

Observe that the three pairs of structures shown in FIG 11 are not transformed to one another by rotations!

Example 4 **The Regular Hexagon**

Construct a generating function of all structures which possess the following general formula

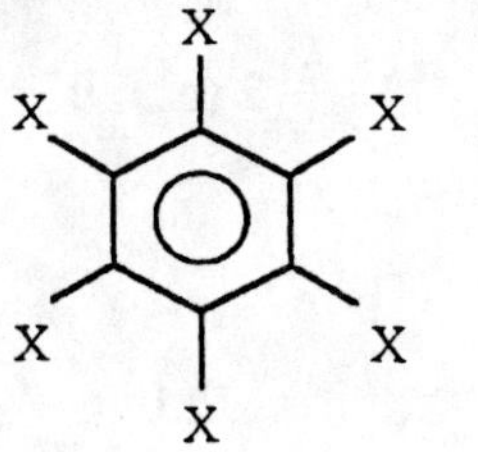

; X = { H, Cl } ... (31)

Here m = 2 , i.e., there are two types only, viz.,

... (32)

The corresponding molecular graph is the regular hexagon. The symmetry operations are shown, together with their permutation representations and S_i^j notations in FIG 13

Figure 13 (see page 163)

The relevant cycle indices are given below:

$$Z_2 \text{ (regular hexagon)} = 6^{-1} [\, S_1^6 + 2\, S_6^1 + S_2^3 + 2\, S_3^2 \,] \qquad \dots (33)$$

$$Z_3 \text{ (regular hexagon)} = 12^{-1} [\, S_1^6 + 2\, S_6^1 + 4\, S_2^3 + 2\, S_3^2 + 3\, S_1^2\, S_2^2 \,] \qquad \dots (34)$$

Then the number of structures in the 2-space is

$$6^{-1} [\, 2^6 + 2.2^1 + 2^3 + 2.2^2 \,] = 13 \quad \text{structures} \qquad \dots (35)$$

On the other hand, in the 3-space one obtains:

$$12^{-1} [\, 2^6 + 2.2^1 + 4.2^3 + 2.2^2 + 3.2^4 \,] = 13 \quad \text{structures} \qquad \dots (36)$$

The "extra" structure can be accounted for by considering the GF in both the 2- and 3-spaces, viz.,

$$\text{GF (2-space)} = a^6 + b^6 + ab^5 + a^5b + 3\, a^2 b^4 + 3\, a^4 b^2 + 4\, a^3 b^3 \dots (37)$$

but, $$\text{GF (3-space)} = a^6 + b^6 + ab^5 + a^5b + 3\, a^2 b^4 + 3\, a^4 b^2 + 3\, a^3 b^3 \dots (38)$$

In 2-space the coefficient of $a^3 b^3$ is 4 , viz.,

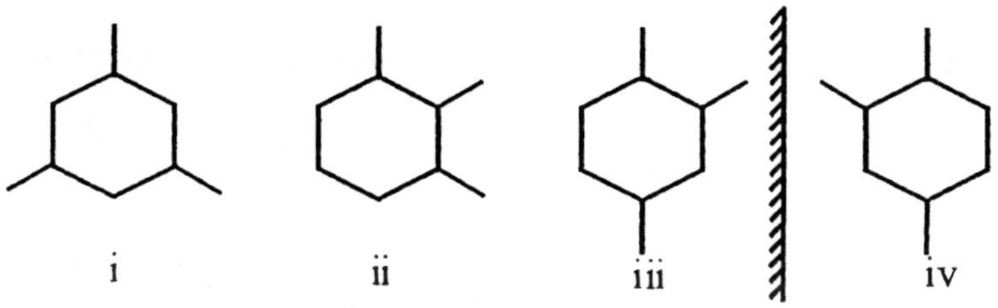

i ii iii iv ... (39)

We observe that iii and iv can be obtained from one another *only via reflection* and *not* by any of the rotations available in Z_2. However the mirror-image of ii can be obtained via 4 rotations successively by 60° each (i.e. by rotation by 240° which is in Z_2). This is portrayed in FIG 14.

Figure 14 (see page 164)

This example demonstrates, once more, that *except when enantiomers are to be counted Z_3 should be used.* Now we consider a somewhat more realistic example, namely, the geometry of the tetrahedron and why there are *two lactic acids*.

Example 5 Enantiomers Of Lactic Acids

Find the generating function of the compounds which possess the general formula given below

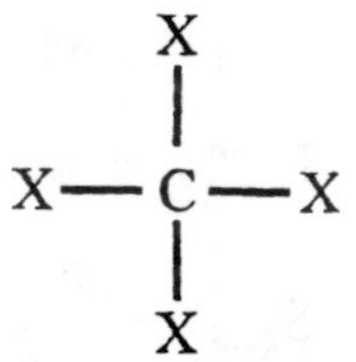

$$X - \underset{X}{\overset{X}{C}} - X \; ; \; X = \{H, CH_3, OH, COOH\} \quad \ldots (40)$$

Use the ideas of Z_2 and Z_3 to conclude that there are in reality two lactic acids, $CH_3\,CH\,(OH)\,COOH$'s.

In fact C atom is not planar, i.e., we can't fit the four valencies of carbon in one plane. However two bonds can be fitted to one plane, one bond projects above the plane and one below the plane. This is the geometry of the tetrahedron which belongs to the $\mathbf{T_d}$ point group. The tetrahedron possesses four C_3 axes, three C_2 axes, three S_4 axes and six σ_d planes. [Observe that the **T** group shown in FIG 2 is a subgroup of the $\mathbf{T_d}$ point group]. Fig 15 portrays the symmetry operations of the tetrahedon.

Figure 15 (see page 165)

The reader may guess the multiplication factors before the cyclic structures which correspond to the various operations, e.g., $8\ S_1^1 S_3^1$ are accounted for as follows: There are four C_3 axes (one axis through each of the four vertices) and there are two types of axes: a C_3 involving clockwise rotation by 120° and a C_3' involving counterclockwise rotation by 120° (= clockwise rotation by 240°), whence the factor of 4 x 2 = 8. Similarly there are three C_2 axes: the one shown in the figure which bisects edges 12 and 34 and one which bisects edges 23 and 14 and a third one passing through edges 13 and 24. In the S_4 operation there are two C_4 axes, one involves a 90° rotation in a clockwise way and the other a 90° rotation in a counter-clockwise manner. Each one bisects two opposite faces (of the cube in which the tetrahedron is embeded, see FIG 15 and since we have 3 pairs of opposite faces, there will be a total of 3 x 2 = 6 S_4 operations. There are six σ_d reflections each one of which fixes two vertices (of the tetrahedron) and permutes the other two vertices. In addition to the one shown, (1)(2)(34), we have five more, viz., (1)(3)(24); (2)(3)(14); (2)(4)(13); (1)(4)(23); and (3)(4)(12). It is remarkable to observe that the "simple" skeleton of methane has 24 symmetry operations! (12 rotations and 12 reflections). The expressions for the cycle indices in 2- and 3-spaces are given below:

$$Z_2\text{ (tetrahedron)} = 12^{-1}\ [S_1^4 + 8\ S_1^1 S_3^1 + 3\ S_2^2\ ] \qquad \dots (41)$$

$$Z_3\text{ (tetrahedron)} = 24^{-1}\ [\ S_1^4 + 8 S_1^1 S_3^1 + 3\ S_2^2 + 6\ S_4^1 + 6\ S_1^2 S_2^1\ ] \qquad \dots (42)$$

Now, m = 4 (because there is one available valency at each C atom and four possible colors, viz., H, CH_3, OH, and CO_2H) and whence in the 2-space one obtains:

$$12^{-1}\ [\ 4^4 + 8.4^2 + 3.4^2\ ] = 36 \text{ structures} \qquad \dots (43)$$

In the 3-space, however, one obtains (using eqn. 42) only 35 structures! This difference an be accounted for by expanding the GF in 2-space, one obtains the following inventory, (where a = CH_3 ; b = H ; C = OH ; d = COOH):

$$\begin{aligned}\text{FG (2-spacer)} = \; & a^4 + b^4 + c^4 + d^4 + \\ & ab^3 + ac^3 + ad^3 + ba^3 + bc^3 + bd^3 + \\ & ca^3 + cb^3 + cd^3 + da^3 + db^3 + dc^3 + \\ & a^2b^2 + a^2c^2 + a^2d^2 + c^2d^2 + b^2d^2 + c^2b^2 + \\ & a^2bc + a^2cd + a^2bd + b^2ac + b^2ad + b^2cd + \\ & c^2ad + c^2ac + c^2bd + d^2ab + d^2ac + d^2bc \\ & + \mathbf{2\,abcd}\end{aligned} \quad \text{... (44)}$$

The last two configuration are counted as one structure in the 3-space. These two configurations are nothing else but the two enantiomers of lactic acid! (viz., *R*-lactic acid and *S*-lactic acid)

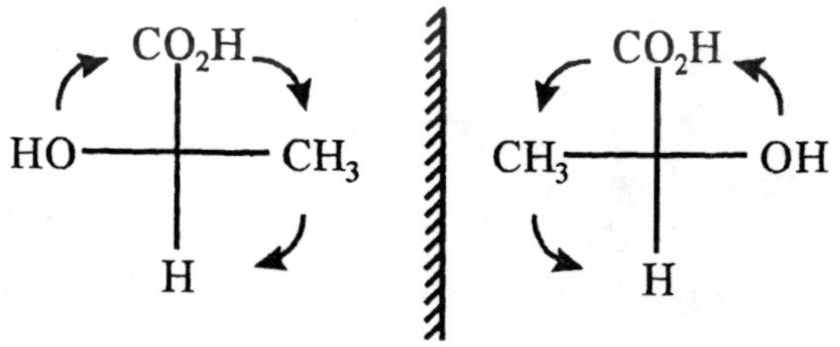

CO_2H / HO — CH_3 / H | CO_2H / CH_3 — OH / H ... (45)

Here (and perhaps for the first time!) an organic chemistry student sees the logic of J.H. van't hoff in concluding that the carbon atom has indeed a tetrahedral configuration.

Sometimes in a given enumeration one is interested in *the number of structures of a particular "weight" only* rather than the entire disply of the pattern inventory. For example, suppose we like to know the number of lactic acids (only) in the present example. In this case we would be interested in the *coefficient* of the term abcd in the relevant GF. The 2- and 3-space – GF's have the following forms:

$$\text{GF (2-space)} = 12^{-1}\,[(a+b+c+d)^4 + 8\,(a+b+c+d)\,(a^3+b^3+c^3+d^3) + 3\,(a^2+b^2+c^2+d^2)\,]\;; \quad \text{... (46)}$$

$$GF\ (3\text{-space}) = 24^{-1}[(a+b+c+d)^4 + 8\,(a+b+c+d)\,(a^3+b^3+c^3+d^3) + 3\,(a^2+b^2+c^2+d^2) + 6\,(a^4+b^4+c^4+d^4) + 6\,(a+b+c+d)^2\,(a^2+b^2+c^2+d^2)] \quad \ldots (47)$$

We then *extract* the required coefficient from the appropriate parentheses. In this case we look for the parentheses which can generate the term abcd. In both GF's only the first one, $(a+b+c+d)^4$ can generate abcd. Then, in the 2-space the coefficient of abcd is:

$$12^{-1}\begin{pmatrix} & & 4 & \\ 1 & 1 & 1 & 1 \end{pmatrix} = 2 \quad \ldots (48)$$

The 2 is the count of the two enantiomers of lactic acid shown in their Fischer projections in eqn. (45).

In the 3-space the operations of reflection count the two enantiomers as one (distinct) structure, i.e. consider the two enantiomers equivalent., viz.,

Number of structures possessing weight = abcd in the 3-space is given by

$$24^{-1}\begin{pmatrix} & 4 & & \\ 1 & 1 & 1 & 1 \end{pmatrix} = 1 \quad \ldots (49)$$

P.S.[1]

$$\begin{pmatrix} n \\ n_1 \quad n_2 \cdots n_k \end{pmatrix} = \frac{n!}{n_1!\ n_2! \cdots n_k!} \quad \ldots (50)$$

Whence we might say that Z_2 "resolves" , so-to-speak, lactic acid into two enantiomess. This "algebraic" resolution will be used whenever there is a possibility of optical isomerism to count enantiomers as we shall see in other examples.

Example 6 Stereoisomerism In Substituted Cyclopropanes [3]

Find the number of disubstituted, trisubstituted and tetrasubstituted enantiomers of cyclopropanes of the general formula

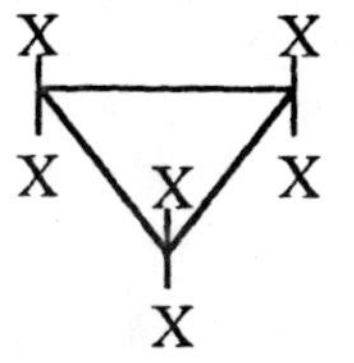

$$; \ X = \{ H, Me \} \qquad \ldots (51)$$

Enantiomers are mirror-images of one-another. Therefore inorder to count enantiomers one includes *only rotations* , i.e., uses Z_2. In FIG 16 we show this (sub-) group of rotations which turns out to be a $\mathbf{D}_3$ subgroup [c.f. FIG 1]. Also, in FIG 16 we show the permutation representations of all symmetry operations.

Figure 16 (see page 166)

The resulting 2-space cycle index is given by the following expression :

$$Z_2 \text{ (Equilateral triangle)} = 6^{-1} \left[S_1^6 + 2\,S_3^2 + 3\,S_2^3 \right] \qquad \ldots (52)$$

We now assume the following weights:

$$X_1 = w\,(Me) = a \ ; \ X_2 = w\,(H) = b \qquad \ldots (53)$$

The generating function in the 2-dimensional space is then given by :

$$GF \text{ (2-space)} = 6^{-1} \left[(a+b)^6 + 2\,(a^3 + b^3)^2 + 3\,(a^2 + b^2)^3 \right] \qquad \ldots (54)$$

Now to find the number of disubstituted derivatives (including enantiomers) we look for the coefficient of $a^2\ b^4$ in the above GF. The following parentheses lead to this coefficient:

$$(a + b)^6 \rightarrow \begin{pmatrix} & 6 & \\ 2 & & 4 \end{pmatrix} = \frac{6\,!}{2\,!\ 4\,!} = 15\ a^2 b^4 \qquad \ldots (55)$$

$$3\,(a^2 + b^2)^3 = 3\,(A + B)^3 \rightarrow 3 \begin{pmatrix} & 3 & \\ 1 & & 2 \end{pmatrix} = 9\,a^2 b^4 \qquad \ldots (56)$$

(where $A = a^2$; $B = b^2$ and we look for coefficient of $a^2 b^4 = A B^2$).

Then we have a total of $(15 + 9) / 6 = 4$ disubstituted derivatives. These are shown in FIG 17 where chiral atoms in pair of enantiomers are marked with asterices.

Figure 17 (see page 167)

The number of trisubstituted isomers (including enantiomers) = the coefficient of a^3b^3. This is calculated from the appropriate parentheses of eqn. (54) as follows :

$$(a + b)^6 \rightarrow \binom{6}{3\ \ 3} = 20\ a^3 b^3 \qquad \text{... (57)}$$

$$2\,(a^3 + b^3)^2 = 2\,(\alpha + \beta)^2 \rightarrow 2 \binom{2}{1\ \ 1} = 4\,a^3b^3 \qquad \text{... (58)}$$

(where $a^3 = \alpha$, $b^3 = \beta$ and $a^3 b^3 = \alpha\beta$)

Then we have $(20 + 4)/6 = 4$ trisubstituted isomers. These are also shown in FIG 17 For the tetrasubstituted isomers, one observes that the coefficient of $a^2 b^4$ = coefficient of $a^4 b^2$ (by symmetry) and therefore there are four tetramethyl isomers.

Example 7 **Stereoisomerism In Substituted Cyclopropanes** [3]

Find the number of geometric *cic-trans* isomers of di- tri- and tetrasubstituted cyclopropanes model in the previous example.

To count geometric isomers *only* one must *exclude* enantiomers by *including* all reflections. I.e., we must add to the group of six rotations shown in FIG 16 six reflection elements [The total of 12 operations, then constitute the $\mathbf{D_{3h}}$ group shown in FIG 2]. In FIG 18 we show the six reflection operations and their permutation representations.

Figure 18 (see page 168)

When we sum the twelve operations (six rotations plus six reflections) we obtain the following cycle index in the 3-space:

$$Z_3\,(\text{triangle}) = 12^{-1}\,[\,S_1^6 + 2\,S_3^2 + 4\,S_2^3 + 3\,S_1^2\,S_2^2 + 2\,S_6^1\,] \qquad \dots (59)$$

The corresponding generating function is given by :

$$\text{GF (3-space)} = 12^{-1}\,[\,(a+b)^6 + 2\,(a^3+b^3)^2 + 4\,(a^2+b^2)^3$$
$$3\,(a+b)^2\,(a^2+b^2)^2 + 2\,(a^6+b^6)\,] \qquad \dots (60)$$

where a and b are defined in eqn. (53).

The number of di- (and tetra-) substituted geometric isomers = the coefficient of $a^2\,b^4$. This coefficient is computed as follows:

$$(a+b)^6 \rightarrow \begin{pmatrix} 6 \\ 2 \quad 4 \end{pmatrix} = \frac{6\,!}{2\,!\;4\,!} = 15\,a^2\,b^4$$

$$4\,(a^2+b^2)^3 \rightarrow 4\,(A+B)^3 \rightarrow 4\begin{pmatrix} 3 \\ 1 \quad 2 \end{pmatrix} = 12\;a^2\,b^4$$

$$(A = a^2\,,\;\; B = b^2\;,\;\; AB^2 = a^2\,b^4\,).$$

$$3(a+b)^2\,(a^2+b^2)^2 = 3\,(\underline{a^2} + \underline{b^2} + 2\,ab)\,(\,a^4 + \underline{b^4} + \underline{2a^2\,b^2}\,) \rightarrow 9\,a^2\,b^4$$

Then the coefficient of $a^2\,b^4 = (15 + 12 + 9) / 12 = 3$

Then there are three *cis-trans* isomers. These are shown in FIG 17.

The number of tri substituted geometric isomers = coefficient of $a^3\,b^3$ ($= \alpha\,\beta$) which is calculated from eqn. (60) as follows:

$$(a+b)^6 \rightarrow \begin{pmatrix} 6 \\ 3 \quad 3 \end{pmatrix} = 20\;a^3\,b^3$$

$$2\,(a^3+b^3)^2 = 2\,(\,\alpha+\beta\,)^2 \rightarrow 2\begin{pmatrix} 2 \\ 1 \quad 1 \end{pmatrix} = 4\;a^3\,b^3$$

Then we have $(20 + 4) / 12 = 3$.

There are therefore three geometric trisubstituted isomers of cyclopropane derivatives. These isomers are also shown in FIG 17.

Example 8 ***cis-trans* Isomers** [4]

Find the number of *cis-trans* isomers of 1,6-dimethyl 1,3,5-hexatriene and draw these isomers.

Here we adopt the following simple model:

$$CH_3\text{-}CH{=}CH\text{-}CH{=}CH\text{-}CH{=}CH\text{-}CH_3 \sim \text{o—o—o} \qquad \dots (61)$$

Where each double bond is represented by a vertex. Then the cycle index of the model (which is a *path* on three vertices) is given by :

$$Z_3\,(\text{o—o—o}) = 2^{-1}\ [\ S_1^3 + S_1^1\,S_2^1\ ] \qquad \dots (62)$$

Where the first term results from application of the identity operation while the second term results from reflection through a plane which passes through the middle vertex. Such a plane fixes the middle vertex to produce a 1-cycle, S_1^1 , and permutes the two terminal vertices to generate a 2-cycle, viz., S_2^1 , whence the net term is $S_1^1\,S_2^1$. There are two possible "colors" for the vertices, one color, say black, represents a *cis* double bond and a white color represents a *trans* double bond. Symbolizing *cic* and *trans* by c and t respectively we have the following GF:

$$GF\,(\text{o—o—o}) = 2^{-1}\ [\ (c+t)^3 + (c+t)\,(c^2+t^2)\ ] \qquad \dots (63)$$

Which when expanded generates the following pattern inventory:

$$c^3 + t^3 + 2\,c\,t^2 + 2\,c^2\,t \qquad \dots (64)$$

I.e. there are six geometric isomers. These are drawn in FIG 19.

Figure 19 (see page 169)

Of course we may also use the equation of Burnside[1,2], eqn. (14), to obtain the number of these isomers: namely, m = 2 (c and t) and we use the expression for Z_3 , eqn. (62) to obtain :

$$\text{Number of geometric isomers} = 2^{-1}\ [\ 2^3 + 2^2\ ] = 6 \qquad \dots (65)$$

Example 9 Axial and Equatorial Bonds[5]

Find the number of mono-, di- and tri-chlorocyclohexanes. Assume a Conventional chair conformation.

In the chair, conformation there are in $2\,C_3 + 3\,C_2 + i + 2\,S_6 + 3\sigma_d$ addition to the identity element. The twelve elements of symmetry constitute the $\mathbf{D_{3d}}$ point group [C.f., FIG 2]. The symmetry operations together with their permutation representations are portrayed in FIG 20.

Figure 20 (see page 170)

This is an example where models are almost essential to understand labelings and whence the cyclic structures of the individual operations. There are 3 ways in which C_2 operates bisecting either one of the 3 pairs of opposite edges. That is why there is a factor of 3 before its cyclic notation. Also, σ_d fixes positions {1, 7, 4, 10} or {2, 8, 5, 11} or { 6, 12, 3, 9}, whence a factor of 3 also before its cyclic structure. There are two ways in which C_6 operates (clockwise and counterclockwise), thus is multiplied by a factor of 2.

The cycle index sums up all 12 operations to give the following expression:

$$Z_3\,(\text{Chair conformation}) = 12^{-1}\,[S_1^{12} + 4S_2^6 + 2\,S_3^4 + 2\,S_6^2 + 3\,S_1^4\,S_2^4\,] \quad \ldots (66)$$

We assume the following weights:

$$W\,(Cl) = a \quad ; \quad w\,(H) = b \qquad \ldots (67)$$

Then the number of monochloro- isomers will be the coefficient ab^{11} in the corresponding GF, viz.,

$$GF = 12^{-1}\,[(a+b)^{12} + 4\,(a^2+b^2)^6 + 2\,(a^3+b^3)^4 + 2\,(a^6+b^6)^2 + 3\,(a+b)^4\,(a^2+b^2)^4\,] \qquad \ldots (68)$$

Only the first and last parentheses generate the required coefficient as follows:

$$(a+b)^{12} \rightarrow \begin{pmatrix} 12 \\ 1 \quad 11 \end{pmatrix} = 12\,a\,b^{11} \quad ;$$

$$3\,(a+b)^4 \quad \times \quad (a^2+b^2)^4$$

$$\downarrow \qquad\qquad \downarrow$$

$$3\begin{pmatrix} & 4 & \\ 1 & & 3 \end{pmatrix} a\,b^3 \quad \times \quad 1\,b^8 \quad \rightarrow \quad 12\,a\,b^{11}$$

Then, there are (12+12)/12 = 2 ab^{11} in the full pattern inventary which correspond to two monochlorocyclohexanes. In one isomer the Cl atom occupies an *equatorial* position and in the other, it occupies an *axial* one. Such an equilibrium is usually illustrated in organic chemistry texts in the following way[5]:

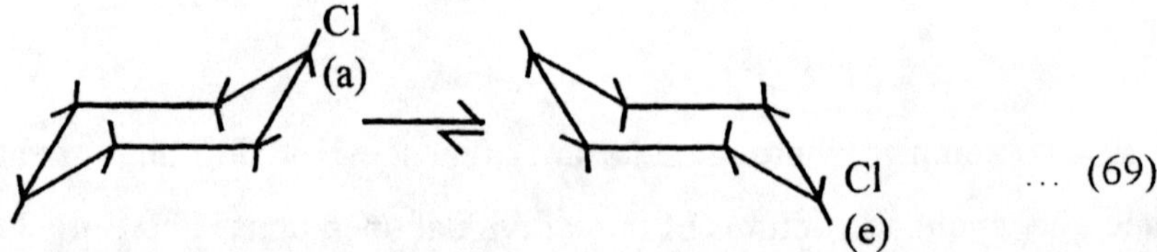

... (69)

where (a) = axial ; (e) = equatorial

Here a reader traces the mathematical origin of the two types of bonds (axial/equatorial) in cyclohexane.

The number of disubstituted isomers is the coefficient of $a^2\,b^{10}$ in the GF given by eqn. (68). We extract the required coefficient as follows:

$$(a+b)^{12} \quad \rightarrow \quad \begin{pmatrix} & 12 & \\ 2 & & 10 \end{pmatrix} = 66\,a^2 b^{10} \,,$$

$$4(a^2+b^2)^6 = 4\,(A+B)^6 \rightarrow 4\begin{pmatrix} & 6 & \\ 1 & & 5 \end{pmatrix} AB^5 = 4 \times 6 = 24\,a^2\,b^{10} \,,$$

$$3\,(a+b)^4 \quad \times \quad (a^2+b^2)^4$$

$$\downarrow \qquad\qquad \downarrow$$

$$3\begin{pmatrix} & 4 & \\ 2 & & 2 \end{pmatrix} a^2\,b^2 \quad \times \quad 1\,b^8 \quad = \quad 18\,a^2\,b^{10} \,,$$

Also;

$$3\,(a+b)^4 \quad \times \quad (a^2+b^2)^4$$

$$\downarrow \qquad\qquad \downarrow$$

$$\begin{pmatrix} 4 \\ 4 \end{pmatrix} \qquad\qquad \begin{pmatrix} & 4 & \\ 1 & & 3 \end{pmatrix}$$

$$3\binom{4}{4} b^4 \quad \times \quad \binom{4}{1\ 3} AB^3 \quad = \quad 12\, a^2 b^{10}$$

Then the number of disubstituted isomers =

$$(66 + 24 + 18 + 12\ ) \div 12 = 10 \qquad \ldots (70)$$

These isomers are abbreviated below (a = axial ; e = equatorial)

1,1 (a e) ;	1,2 (a a) ;	1,2 (a e) ;	1,2 (e e)
1,3 (a a) ;	1,3 (a e) ;	1,3 (e e) ;	
1,4 (a a) ;	1,4 (a e) ;	1,4 (e e)	

... (71)

The number of trisubstituted isomers is the coefficient of $a^3 b^9$ in the GF (eqn. 68). This is calcualted as follows:

$$(a + b)^{12} \rightarrow \binom{12}{3\ 9} = 220\, a^3 b^9 ,$$

$$2(a^3 + b^3)^4 = 2(\alpha + \beta)^4 \rightarrow 2\binom{4}{1\ 3} \alpha\beta^3 = 8\, a^3 b^9 ;$$

$$\begin{array}{lccl} 3\,(a+b)^4 & \times & (a^2+b^2)^4 & \\ \downarrow & & \downarrow & \\ 3 \times 4\, a^3 b & \times & b^8 & \rightarrow 12\, a^3 b^9 \\ 3 \times 4\, a\, b^3 & \times & 4\, a^2 b^6 & \rightarrow 48\, a^3 b^9 \end{array}$$

Then the number of trichlorocyclohexanes = (220+8+12+48) ÷ 12

= 24 ... (72)

These 24 isomers are symbolized below:

	a	a	a		
	e	e	e		
Positions 1 2 3	a	e	e	→	6 isomers
	a	a	e		
	a	e	a		
	e	a	e		

There are also 6 isomers of the above type at positions 1 2 4 and 1 3 5

In addition six more types exist as follows:

1 (e) , 1 (a) , 2 (e)
1 (e) , 1 (a) , 2 (a)
1 (e) , 1 (a) , 3 (e)
1 (e) , 1 (a) , 3 (a)
1 (e) , 1 (a) , 4 (e)
1 (e) , 1 (a) , 4 (a)

Then the total number is $6 \times 3 + 6 = 24$ ... (73)

Observe that substitution at positions 1,2,4 is equivalent to that at positions 1,3,4, i.e.,

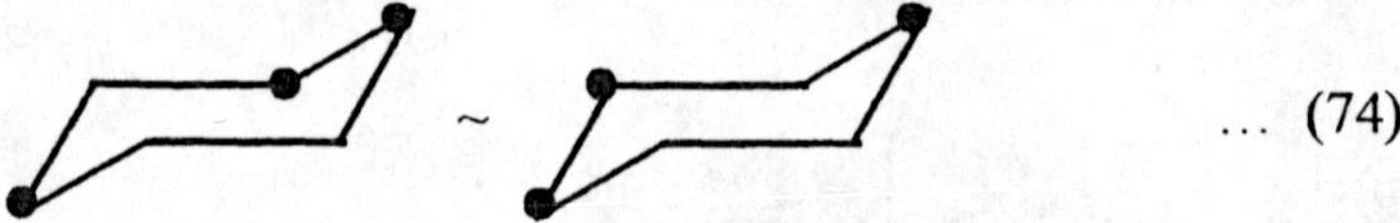
... (74)

Example 10 **Stereoisomerism of Cyclobutanes** [3]

Find the number of diastereomers of dichlorocyclobutanes and draw these isomers.

In this case the possibility of enantiomers exist and therefore we compute isomer count both in 2- and 3-spaces. There are 16 symmetry operations forming altogether the $\mathbf{D}_{4h}$ group. These operations are illustrated in FIG 21

Figure 21 (see page 171)

The expression for Z_2 (rotations only) and Z_3 (rotations and reflections) are given below:

$$Z_2 = 8^{-1} [S_1^8 + 2 S_4^2 + 5 S_2^4] \quad \ldots (75)$$

$$Z_3 = 16^{-1} [S_1^8 + 4S_4^2 + 9S_2^4 + 2S_1^4 S_2^2] \quad \ldots (76)$$

The GF which corresponds to Z_2 generates all diastereomers *including enantiomers* but the GF which belongs to Z_3 generates geometrical isomers while pairs of enantiomers are counted as ones. We use the following weights:

$$W(H) = a \quad ; \quad w(Cl) = b \qquad \ldots (77)$$

Then,

$$GF(2\text{-space}) = 8^{-1}\,[(a+b)^8 + 2\,(a^4+b^4)^2 + 5\,(a^2+b^2)^4] \qquad \ldots (78)$$

The number of dichlorocyclobutanes (including enantiomers) is the coefficient of a^6b^2(= coefficient of a^2b^6), This is computed as follows:

$$(a+b)^8 \rightarrow \binom{8}{6\ \ 2} = 28\ a^6\,b^2$$

$$5\,(a^2+b^2)^4 = 5\,(A+B)^4 \rightarrow 5 \times \binom{4}{3\ \ 1} = 20\,A^3\,B = 20\,a^6\,b^2$$

Whence the number of diastereomers including enantiomers = (28 + 20) / 8 = 6 isomers. The six compounds are shown in FIG 22.

Figure 22 (see page 172)

Observe that isomers 1-4 all contain an element of symmetry (a plane or an inversion center) and whence are counted only once (even using Z_2) . However isomers 5 and 6 contain no elements of symmetry (except the identity !).

Now when we use Z_3 (i.e. include reflections) enaniomers 5 and 6 will be counted as one compound and we should expect only 5 isomers. These are calculated as follows:

$$GF(3\text{-space}) = 10^{-1}\,[(a+b)^8 + 4(a^4+b^4)^2 + 9\,(a^2+b^2)^4 + 2\,(a+b)^4\,(a^2+b^2)^2\,] \qquad \ldots (79)$$

The coefficient of $a^6\,b^2$ is extracted from the relevant parentheses as follows

$$(a+b)^8 \rightarrow 28$$

$$9\,(a^2+b^2)^4 \rightarrow 36$$

$$2\,(a+b)^4\,(a^2+b^2)^2 \;\rightarrow\; 16$$

whence we have $(28 + 36 + 16) / 16 = 5$ isomers.

Example 11 **Stereochemistry of Allene** [6]

Find all dichloroallenes including stereoisomers.

Allene has two double bonds whose p-orbitals are perpendicular to each other. It belongs to the $\mathbf{D}_{2d}$ point group modeled in FIG 2 . One way to envisage the molecule of allene is to inscribe it into an "elongated" cube as shown in FIG 23 which also shows a top-view of this molecule.

Figure 23 (see page 173)

Allene possesses the following symmetry operations (labelings correspond to FIG 23)

$I \rightarrow (1)(2)(3)(4) \rightarrow S_1^4$

$C_2 \rightarrow (12)(34) \rightarrow S_2^2$

$C_2' \rightarrow (14)(23) \rightarrow S_2^2$

$C_2'' \rightarrow (13)(24) \rightarrow S_2^2$

$\sigma \rightarrow (1)(2)(34) \rightarrow S_1^2 S_2^1$

$\sigma' \rightarrow (3)(4)(12) \rightarrow S_1^2 S_2^1$

$S_4 \rightarrow (1\,4\,2\,3) \rightarrow S_4$

$S_4' \rightarrow (1\,3\,2\,4) \rightarrow S_4$

The S_4 and S_4' operations are illustrated in FIG 23 .

The four rotations lead to Z_2 :

$$Z_4 = 4^{-1} \; [\, S_1^4 + 3\, S_2^2 \,] \qquad \ldots (80)$$

while the four rotations and four reflections lead to Z_3 :

To find the number of isomers including enantiomers we use Z_2 to form GF, viz.,

$$\text{GF (2-space)} = 4^{-1} \, [(a+b)^4 \; + \; 3\,(a^2+b^2\,)^2\,] \qquad \ldots (81)$$

Where w(H) = a, w (Cl) = b . Then the number of dichloroisomers including optical isomers is the coefficient of a^2b^2 which is obtained from eqn (81) as follows:

$$(a+b\,)^4 \;\rightarrow \begin{pmatrix} & 4 & \\ 2 & & 2 \end{pmatrix} \quad = 6\; a^2\, b^2 \;;$$

$$3\,(a^2+b^2)^2 \;\;\rightarrow 3 \begin{pmatrix} & 2 & \\ 1 & & 1 \end{pmatrix} = \; 6\, a^2\, b^2$$

Whence, the number of isomers including enantiomers = (6+6)/4 = 3. The three isomers are shown in FIG 24 where the pair of enantiomers are seperated by a mirror plane.

Figure 24 (see page 174)

In the 3-space reflections enter into the picture and we expect the two optical isomers to be counted as a single compound (because Z_3 "sees" them as "identical" because one can be obtained from the other by reflection in a mirror plane). Indeed this turns out to be the case, namely, we have:

$$\text{GF (3-space)} = 8^{-1} \, [\, (a+b)^4 \; + \; 3\,(a^2+b^2\,)^{\,2} \; + \; 2\,(a^4+b^4) \\ + 2(a+b)^2\,(a^2+b^2)\,] \qquad \ldots (82)$$

The coefficient of $a^2\, b^2$ is calculated as follows:

$(a+b)^4 \;\rightarrow 6\; a^2\, b^2$; $3\,(a^2+b^2\,)^2 \;\rightarrow\; 6\, a^2\, b^2$; $2(a+b)^2\,(a^2+b^2) \rightarrow$ $4\; a^2\; b^2$, and therefore the coefficient of $a^2\, b^2$ = (6+6+4) / 8 = 2 = the number of isomers not including enantiomers. (In this case the two isomers are related as diastereomers).

Example 12 **Stereochemistry of Cubane** [7]

Methyl cubanes are known to be stable compounds. Find all di- and trimethyl cubanes.

The geometry of the cube, its symmetry operations and their permutation representations are all shown in FIG 25. which represent the **O** point group.

Figure 25 (see page 175)

Observe in FIG 25 that operation types a and b involve opposite faces, operation c involves opposite edges while operation d involves opposite vertices. From the cyclic structures shown in FIG 25 one can easily write the expression for the generating function, viz.,

$$GF = 24^{-1} [(a+b)^8 + 6 (a^4 + b^4)^2 + 9 (a^2 + b^2)^4 + 8 (a+b)^2 (a^3 + b^3)^2] \quad \ldots (83)$$

where $a = w(Me)$; $b = w(H)$

Using multinomial theorem we obtain the following results:

The number of dimethyl cubanes = coefficient of $a^2 b^6 = 3$;

The number of trimethyl cubanes = coefficient of $a^3 b^5 = 3$.

The relevant isomers are drawn in FIG 26.

Figure 26 (see page 176)

Example 13 **Stereochemistry of the Octahedron** [8]

Find the number of octahedral isomers of the general formulas: $Pt\,Br_4\,Cl_2$; $Ru\,Cl_4\,(H_2O)\,(CO)$; and $Ru\,Cl\,I\,(CH_3)\,CO\,(PPh_3)_2$.

Here we only need the 24 symmetry operations which belong to the **O** group [C.f. previous example]. FIG 27 illustrates the symmetry celements of the octahedron. The symmetry operations are outlined below : (The labeling are those of FIG 27).

$$I \rightarrow S_1^6$$

$$6\ C_4 \rightarrow \text{e.g., } (5)\,(6)\,(1234) \rightarrow 6\ S_1^2\, S_4^1$$

$$3\ C_2 \rightarrow \text{e.g., } (5)\,(6)\,(24)(13)) \rightarrow 3\ S_1^2\, S_2^2$$

$$6\ C_2 \rightarrow \text{e.g., } (12)\,(34)\,(56) \rightarrow 6\ S_2^3$$

$$8\ C_3 \rightarrow \text{e.g., } (145)\,(263) \rightarrow 6\ S_3^2$$

Figure 27 (see page 177)

The corresponding GF is given by:

$$GF = 24^{-1}\,[\,(w_1 + w_2 + .. + w_6)^6 + 6\,(w_1 + w_2 + .. + w_6)^2\,(w_1^4 + w_2^4 + \ldots + w_6^4) +$$
$$3\,(w_1 + w_2 + .. + w_6)^2\,(w_1^2 + w_2^2 + \ldots + w_6^2)^2 + 8\,(w_1^3 + w_2^3 + \ldots + w_6^3)^2$$
$$+ 6\,(w_1^2 + w_2^2 + .. + w_6^2)^3\,] \qquad \ldots (84)$$

where the w`s are weights of individual substituents.

Conventional expansion of the above GF leads to the desired results, viz.,

- Coefficient of $w_1^4\, w_2^2$ = number of isomers of Pt Br_4 Cl_2 = 2
- Coefficient of $w_1^4\ w_2\, w_3$ = number of isomers of Ru Cl_4 $(H_2O)CO$ = 2
- Coefficient of $w_1\ w_2\ w_3\ w_4\ w_5^2$ = number of isomers of Ru Cl I (CH_3) CO $(Ph_3)_2$ = 15.

2.5.2 Combinatorial Generation of Heterocycles

Example 14 **Combinatorial Generation of Heterocycles**

Find the number of heterocycles of the type shown below:

X
X X ; X= {C, N} ... (85)
X——X

which contain 2N atoms. Draw these heterocycles.

The symmetry operations of the regular pentagon are illustrated in FIG 28.

Figure 28 (see page 178)

Now we let w(C) = a ; w(N) = b , then desired number is the coefficient of $a^3 b^2$ in the following function:

$$GF = 10^{-1} [(a + b)^5 + 4 (a^5 + b^5) + 5 (a + b) (a^2 + b^2)^2] \quad \text{... (86)}$$

Conventional expansion leads to a coefficient of 2 for the term $a^3\ b^2$. The two heterocycles are shown below

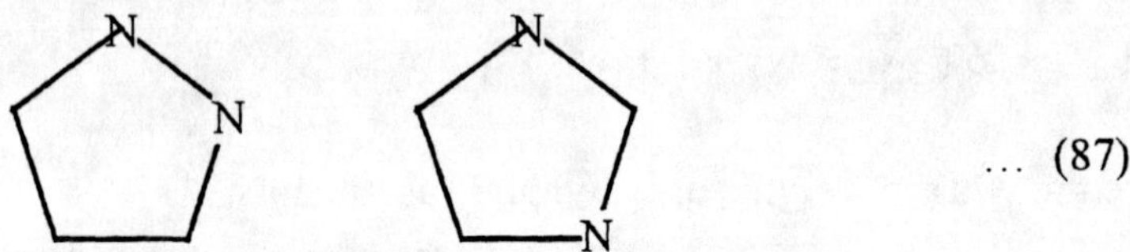

... (87)

Example 15 **Combinatorial Generation of Heterocycles**

Find the heterocycles derived from the regular pentagon, eqn. (85) when X= {C, N, S} which contain 1N and 1S atom.

We assume the folowing weights:

W(C) = a ; w(N) = b ; w(S) = c

Then we look for the coefficient of a^3 bc in the following generating function :

$$GF = 10^{-1}\,[(a+b+c)^5 + 4(a^5+b^5+c^5) + 5(a+b+c)\,(a^2+b^2+c^2)^2] \qquad \ldots (88)$$

The required coefficient is 2 , corresponding to :

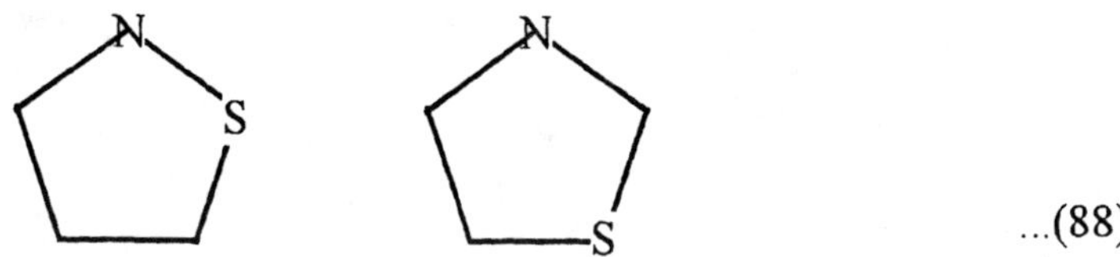

...(88)

Example 16 **Combinatorial Generation of Heterocycles**

Find all heterocycles which can be generated from the regular pentagon where $X = \{C, N, S\}$ and which contain 2N and 1S atoms.

Here we look for the coefficient of $a^2\, b^2\, c$ in eqn. (88) and usual expansion leads to a value of 4 corresponding to the following heterocycles:

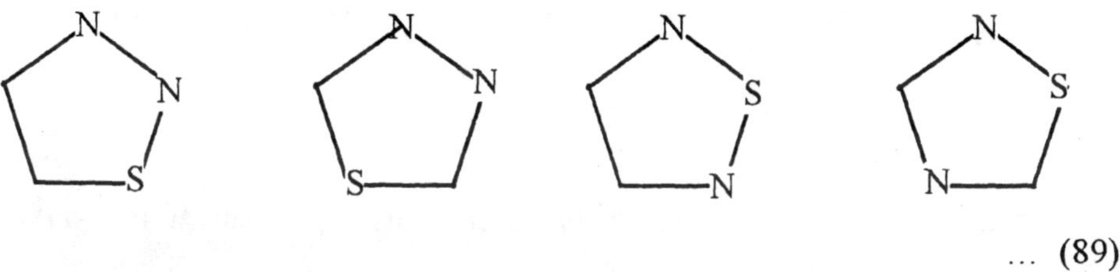

... (89)

Example 17 **Combinatorial Generation of Heterocycles**

Find the heterocycles which contain 2N and 3N atoms which can be derived from the following system

```
    X
 X     X
 |     |     ;  X = {C, N}        ... (90)
 X     X
    X
```

Fig. 29 shows all six rotations and six reflections of the regular hexagon from which one can derive the 3-space cycle index.

Figure 29 (see page 179)

Now,, as usual, $w(C) = a$; $w(N) = b$ leads to the following generating function:

$$GF = 12^{-1} \left[(a+b)^6 + 2(a^6+b^6) + 2(a^3+b^3) + 4(a^2+b^2)^3 + 3(a+b)^2(a^2+b^2) \right] \quad \ldots (91)$$

From which we obtain:

The coefficient of $a^4 b^2 = 3$,

The coefficient of $a^3 b^3 = 3$

These two findings are consistent with three heterocycles in both cases, viz.,

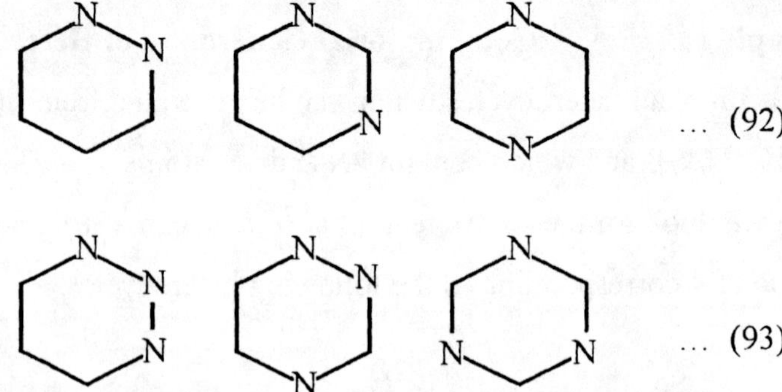

... (92)

... (93)

Example 18 Combinatorial Generation of Heterocycles

Find all heterocycles derived from the previous system shwon in eqn.(90) for which X = {C, N,S} which have 2N and 1S atoms.

Again using w(C) = a , w(N) = b , w(S) = c generates the following function:

$$GF = 12^{-1} \left[(a+b+c)^6 + 2(a^6+b^6+c^6) + 2(a^3+b^3+c^3)^2 + 4(a^2+b^2+c^2)^3 + 3(a+b+c)^2(a^2+b^2+c^2)^2 \right] \quad \ldots (94)$$

The required coefficient is $a^3 b^2 c = 6$ and the 6 heterocycles are shown in FIG 30.

Figure 30 (see page 180)

Example 19 Combinatorial Generation of Heterocycles

Work out the previous system and find all heterocycles possessing 2N and 2S atoms.

In this case we look for the coefficient of $a^2\ b^2\ c^2$ in eqn. (94). Straight-forward algebra leads to a value of 11 for this coefficient. These 11 heterocycles are also shown in FIG 30.

Example 20 Combinatorial Generation of Hetrerocycles

Consider the following general bicyclic system:

```
X——X——X
|  |  |
X——X——X
```

Where : X = {C, N} ; ii) X = { C, N, S }

Find all heterocycles which contain:

1) 2N , 4C atoms 2) 3 N, 3C atoms.

3) 2N , 1 S and 3 C atoms 4) 3N, 1S and 2C atoms

The parent graph is shown in FIG 2 as a model for the C_{2v} point group. The 3-space cycle index can easily be derived: one obtains :

$$Z_3 = 4^{-1} [\ S_1^6 + 2\ S_2^3 + S_1^2\ S_2^2\ ] \qquad \dots (95)$$

Now we let w(C) = a , w(N) = b, w(S) = c , then 1) and 2) lead respectively to the following GF's:

$$GF(1) = 4^{-1} [\ (a+b)^6 + 2\ (a^2 + b^2)^3 + (a+b)^2\ (a^2 + b^2)^2\ ] \qquad \dots (96)$$

$$GF(2) = 4^{-1} [(a+b+c)^6 + 2(a^2+b^2+c^2)^3 + (a+b+c)^2\ (a^2+b^2+c^2)^2] \qquad \dots (97)$$

<u>2N atoms:</u>

The coefficient of $a^4\ b^2$ in eqn. (96) = 6

<u>3N atoms:</u>

The coefficient of $a^3\ b^3$ in eqn. (96) = 6

<u>2N, 1S atoms:</u>

The coefficient of $a^3\ b^2\ c$ in eqn. (97)= 16

<u>3N, 1S atoms</u>

The coefficient of $a^2\ b^3\ c$ in eqn. (97) = 15

Fig 31 shows how to derive the above heterocycles.

Figure 31 (see page 181)

We can go on generate other types of heterocycles using theorem of Pólya. In Table 2 we outline some results of two more bicyclic systems.

Table 2

Number of heterocycles of some bicyclic systems obtained using the theorem of Polya. All systems belong to the $\mathbf{C_{2v}}$ point group.

	System	Atom weights	Number of heterocycles
(1)	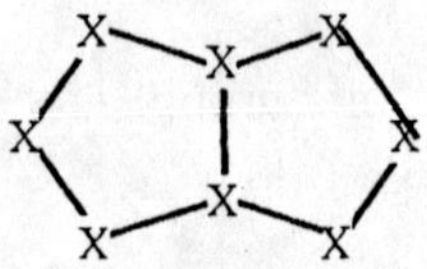 X={C,N}	2N	10
(2)	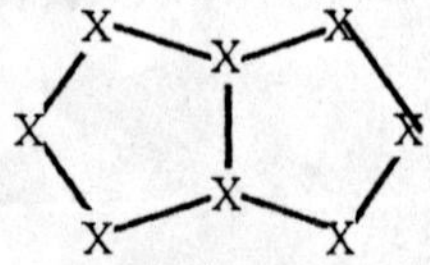 X={C,N,S}	2N, 1S	45
(3)	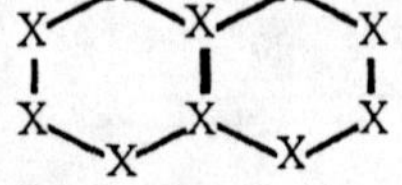 X={C,N}	2N	15
(4)	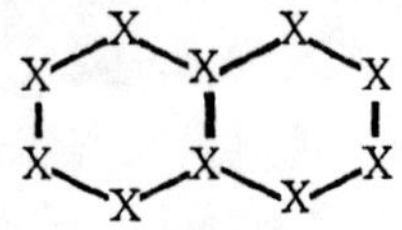 X={C,N,S}	2N, 1S	92

As an illustration we show all 15 heterocycles for system (3) in FIG 32

Figure 32 (see page 182)

2.5.3 Application of Theorem of Pólya in the Study of nmr: Types of H atoms, ^{13}C nmr [9]

Balasubramanian[9] used theorem of Pólya to study low resolution nmr spectra (both pmr, i.e. proton magnetic resonance and ^{13}C nmr) where he computed the number of magnetically equivalent nuclei (protons or C atoms). He used the symbols α_1 , α_2 ... to indicate weights of atoms labeled 1,2,..., then using a rigorous argument he concluded the following relation:

Number of signals = number of magnetically equivalent nuclei = coefficient of the term $\alpha_1 \alpha_2^{n-1}$ (coefficient of $\alpha_1^{n-1} \alpha_2$) in the corresponding generating function.

... (98)

Where n = number of nuclei. By the number of signals we mean those which appear in the very low resolution (zero-order) nmr spectrum.

Here we will not reproduce Balasubramanian's argument which led him to eqn. (98) but will give a "simple-minded" explanation of it : Consider, e.g. pmr: In the lowest possible resolution it enumerates several signals, each one corresponds to a set of magnetically equivalent nuclei (= i.e., to a particular *orbit* of nuclei), but the number of (magnetically) equivalent H atoms, corresponds to the number of mono-substituted derivatives. For example there is only one set of equivalnet H atoms in benzene and indeed there is only one chlorobenzene!. There are two sets of equivalent H atoms in naphthalene (those in the α-position and those in the β position) and infact there are two chloronaphthalenes, and so on. Now if we let (w(H) = a; w(Cl) = b, then the number of (mono)-chloronaphthalens is the coefficient of $a^7 b$ (or of $a b^7$) in the GF which corresponds to the 8H atoms making the periphery of naphthalene molecule. I.e., it is the coefficient of $a^{n-1}b$ in this case. And indeed we expect the zero-order pmr of naphthalene to be composed of two signals. If we replace a and b by α_1 and α_2 we can easily understand the logic of the formula of Balasubramanian, eqn. (98). Analogously, and using only "chemical intuition", one may conclude that there are only three aza hetrocycles derived by replacement of one of the 10 C atoms of naphthalene by an N atom,viz.,

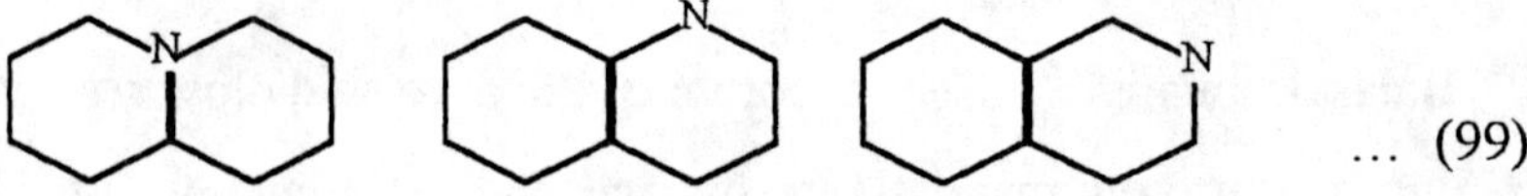

... (99)

Indeed the low resolution ^{13}C nmr of naphthalene tells us that there are 3 sets of (magnetically) equivalent C atoms. Indeed the coefficient of a^9 b (or $\alpha_1^9 \alpha_2$) in GF(C skeleton of naphthalene) = 3 . (Observe that n = 10 nuclei in this case).

We now outline the steps to obtain the number magnetically equivalent sets of nuclei in a given system (= the number of orbits) = the number of zero order nmr signals.

1- Draw the relevant structure and label all nuclei (H atoms for pmr or C atoms for ^{13}C nmr) as 1,2,.., n.

2- Define the point group to which your structure belongs, ie., identify all symmetry operations.

3- Find the expression for Z_3 , the cycle index in the 3-space.

4- Assign weights α_1 , α_2 .., α_n to the n nuclei and find the GF.

5- Extract the coefficient of α_1^{n-1} α_2 (or of $\alpha_1 \alpha_2^{n-1}$) from the GF. This coefficient is the required number of orbits (i.e. number of equivalent nuclei).

2.5.3.1 Examples dealing with pmr.

In FIG 33 we show six systems with an arbitrary labelings of the H atoms. Some of these systems may not be real compounds but they serve to illustrate the procedure.

Figure 33 (see page 183)

Example 21 Computation of pmr

Work out pmr spectrum of system 1 of FIG 33

This is a C_{2v} point group system. The four symmetry operations and their permutation representations are given below:

$$I \rightarrow (1)(2)(3)(4) \rightarrow S_1^4$$

$$C_2 \rightarrow (12)(34) \rightarrow S_2^2$$

$$\sigma_h \rightarrow (3)(4)(12) \rightarrow S_1^2 S_2^1$$

$$\sigma_{12} \rightarrow (1)(2)(34) \rightarrow S_1^2 S_2^1$$

Then, $Z_3 = 4^{-1} [S_1^4 + 2 S_1^2 S_2^1 + S_2^2]$; ... (100)

$$GF = 4^{-1} [(\alpha_1 + \alpha_2 + \ldots)^4 + 2 (\alpha_1 + \alpha_2 + ..)^2 (\alpha_1^2 + \alpha_2^2 + ..) + (\alpha_1^2 + \alpha_2^2 + ..)^2] \quad \ldots (101)$$

The coefficient of $\alpha_1 \alpha_2^3$ is extracted below :

$$(\alpha_1 + \alpha_2 + \ldots)^4 \rightarrow \binom{4}{1\ 3} \rightarrow 4\ \alpha_1 \alpha_2^3$$

$$2 (\alpha_1 + \alpha_2 + \ldots)^2 (\alpha_1^2 + \alpha_2^2 + ..) = 2 (\alpha_1^2 + \alpha_2^2 + 2 \alpha_1 \alpha_2 + ..) (\alpha_1^2 + \alpha_2^2 + ..)$$

$$\rightarrow 4 \alpha_1 \alpha_2^3$$

Then the coefficient of $\alpha_1 \alpha_2^3 = (4 + 4) / 4 = 2$.

Indeed we have two orbits :

$\Delta_1 = \{ 1, 2\}$; $\Delta_2 = \{ 3, 4\}$, See FIG 33

This is an almost trivial case which is considered only to exemplify the method.

Example 22 Computation of pmr

Find the number of sets of magnetically equivalent H atoms of system $\underset{\sim}{2}$ of FIG 33

The six H atoms of this system belong to the $\mathbf{D}_{2h}$ point group. We have the following permutations:

$$I \rightarrow (1)(2)\ldots(6) \rightarrow S_1^6$$

$$C_4^2 \rightarrow (13)(24)(56) \rightarrow S_2^3$$

$$C_2 \rightarrow (5)(6)(14)(23) \rightarrow S_1^2 S_2^2$$

$$C_2' \rightarrow (12)(34)(56) \rightarrow S_2^3$$

$$\sigma_{56} \rightarrow (5)(6)(13)(24) \rightarrow S_1^2 S_2^2$$

$$\sigma_{13} \rightarrow (1)(2)(3)(4)(56) \rightarrow S_1^4 S_2^1$$

$$\sigma_h \text{ (paper plane)} \rightarrow (12)(34)(5)(6) \rightarrow S_1^2 S_2^2$$

$$i \rightarrow (14)(23)(56) \rightarrow S_2^3$$

$$\therefore \quad Z_3 = 8^{-1}\,[S_1^6 + 3\,S_2^3 + 3\,S_1^2 S_2^2 + S_1^4 S_2^1\,]; \qquad \ldots (102)$$

$$GF = 8^{-1}\,[\,(\alpha_1 + \alpha_2 + \ldots)^6 + 3\,(\alpha_1^2 + \alpha_2^2 + \ldots)^3 + 3\,(\alpha_1 + \alpha_2 + \ldots)^2\,(\alpha_1^2 + \alpha_2^2 + \ldots)^2 + (\alpha_1 + \alpha_2 + \ldots)^4\,(\alpha_1^2 + \alpha_2^2 + \ldots)\,] \qquad \ldots (103)$$

The coefficient of $\alpha_1^5\,\alpha_2 = 2$ which leads to two sets of equivalent protons, viz , $\Delta_1 = \{1,2,3,4\}$ and $\Delta_2 = \{5,6\}$. These two sets of H atoms make the two orbits

Δ_1 and Δ_2.

Eample 23 **Computation of pmr**

Find the number of orbits of the 10 H atoms of system $\underset{\sim}{3}$ of FIG 33

Again this is a $\mathbf{D}_{2h}$ system with the following symmetry operations:

$I \rightarrow (1)(2)\ldots(10) \rightarrow S_1^{10}$

$C_4^2 \rightarrow (17)(28)(35)(46)(910) \rightarrow S_2^5$

$C_2 \rightarrow (14)(23)(58)(67)(910) \rightarrow S_2^5$

$C_2' \rightarrow (16)(25)(38)(47)(9)(10) \rightarrow S_1^2 S_2^4$

$\sigma_h \rightarrow (13)(24)(57)(68)(910) \rightarrow S_2^5$

$\sigma_v \rightarrow (15)(26)(37)(48)(9)(10) \rightarrow S_1^2 S_2^4$

σ_h' (paper plane) $\rightarrow (12)(34)(56)(78)(9)(10) \rightarrow S_1^2 S_2^4$

$i \rightarrow (18)(27)(36)(45)(910) \rightarrow S_2^5$

Whence, $Z_3 = 8^{-1} [S_1^{10} + 4 S_2^5 + 3 S_1^2 S_2^4]$ ... (104)

The GF leads to a coefficient $\alpha_1^9 \alpha_1 = 2$ and indeed there are two types of protons: $\Delta_1 = \{ 1, 2, 3, 4, 5, 6, 7, 8\}$ and $\Delta_2 = \{ 9, 10\}$.

Example 24 **Computation of pmr**

Work out system $\underset{\sim}{4}$ of FIG 33 and find out the number of orbits of H atoms. Also, this system belongs to the $\mathbf{D}_{2h}$ point group with the following elements: { I, C_4^2, $C_2, C_2', \sigma_h, \sigma_v, \sigma_h'$ (paper plane), i }

The resulting $Z_3 = 8^{-1} [S_1^8 + 7 S_2^4]$ and the coefficient of $\alpha_1 \alpha_2^7 = 1$ in the corresponding GF which is consistent with only one type of H atoms. I.e., there is just one orbit, $\Delta_1 = \{1,2, \ldots,8 \}$.

Example 25 Computation of pmr

Find the orbits of the H atoms of naphthalene, system $\underset{\sim}{5}$, FIG 33.

This is a C_{2v} point-group system whose $Z_3 = 4^{-1} [S_1^8 + 3 S_2^4]$. The corresponding GF calculates $2 \alpha_1^7 \alpha_2$ and indeed the H atoms of naphthalene generate two orbtis, viz, $\Delta_1 = \{1, 4, 5, 8\}$ and $\Delta_2 = \{2, 3, 6, 7\}$.

Example 26 Computation of pmr

Find the number of types of H atoms in the molecule of anthracene, system $\underset{\sim}{6}$. FIG 33.

This is also a C_{2v} system, the Z_3 of which is $4^{-1} [S_1^{10} + 2 S_2^5 + S_1^2 S_2^4]$. The resulting GF generates $3 \alpha_1^9 \alpha_2$ which is consistent with three types (i.e. three orbits), viz.,

$$\Delta_1 = \{1, 4, 6, 9\} \; ; \; \Delta_2 = \{2, 3, 7, 8\} \; ; \; \Delta_3 = \{5, 10\}.$$

We remined the reader that in all such cases, the number of orbits of H atoms = the number of monosubstituted isomers = the number of types of H atoms = the number of zero-order pmr signals.

2.5.3.2. Examples dealing with ^{13}C nmr

In FIG 34 we show four H-suppressed molecular graphs of four benzenoid systems. The types of nuclei are "colored" differently so that the number of orbits (≡ number of ^{13}C nmr peaks) can be compared with those calculated using theorem of Polya and the equation of Balasubramanian, eqn. (98).

Figure 34 (see page 184)

We consider the following cases, (C.f. FIG 34).

Example 27 : <u>Naphthalene</u> $\underset{\sim}{1}$

$$Z_3 = 4^{-1} [S_1^{10} + 2 S_2^5 + S_1^2 S_2^4] \qquad \text{... (105)}$$

The coefficient of $\alpha_1 \alpha_2^9$ in the GF = 3 which corresponds to three magnetically equivalent nuclei, viz, $\Delta_1 = \{1, 4, 6, 9\}$; $\Delta_2 = \{2, 3, 7, 8\}$; $\Delta_3 = \{5, 10\}$. In FIG 34 Δ_1 is colored as triangles, Δ_2 as open circles and Δ_3 as solid circles.

Example 28 Anthracene $\underset{\sim}{2}$

$$Z_3 = 4^{-1} [S_1^{14} + 2 S_2^7 + S_1^2 S_2^6] \quad \ldots (106)$$

The coefficient of $\alpha_1 \alpha_2^{13}$ in the relevant GF = 4 and the 14 C nuclei Generate the following four orbits (or types):

$\Delta_1 = \{2, 3, 9, 10\}$, colored as open circles.

$\Delta_2 = \{1, 4, 8, 11\}$, colored as solid circles

$\Delta_3 = \{7, 12, 14, 5\}$, colored as open triangles

$\Delta_4 = \{6, 13\}$, colored as solid triangles.

Example 29 Pyrene $\underset{\sim}{3}$

$$Z_3 = 4^{-1} [S_1^{16} + 2 S_2^8 + S_1^4 S_2^6] \quad \ldots (107)$$

The coefficient of $\alpha_1 \alpha_2^{15} = 5$ and there are five orbits, viz.,

$\Delta_1 = \{1, 8, \}$, colored as open circles.

$\Delta_2 = \{2, 7, 9, 14\}$, colored as closed circles

$\Delta_3 = \{3, 6, 10, 13\}$, colored as open triangles

$\Delta_4 = \{4, 5, 11, 12\}$, colored as solid triangles.

$\Delta_5 = \{15, 16, \}$, colored as open squares.

We can, then, conclude immediately that there are 5 positional monochloropyrene isomers !

Example 30 Coronene $\underset{\sim}{4}$

There are six rotations (around an axis perpendicular to molecular plane) by angles of 60°, 120°, 180°, 240°, 300° and 360° and six reflections. (Three σ's bisecting opposite edges and three σ's bisecting opposite vertices). The resulting Z_3 is given by :

$$Z_3 = 12^{-1} [S_1^{24} + 2S_6^4 + 2S_3^8 + 4S_2^{12} + 3S_1^4 S_2^{10}] \qquad \dots (108)$$

The coefficient of $\alpha_1 \alpha_2^{23} = 3$ which is consistent with 3 orbits of equivalent nuclei. These 3 orbits can be easily envisaged from FIG 34. They are colored as open circles, solid squares and solid triangles.

Bibliography

1. D.I. A. Cohen Basic techniques of combinatorial theory, John Wiley & Sons, New York, Chapter 6 (1978).
2. R.P. Grimaldi, Discrete and combinatorial mathematics, an applied introduction, Addison-Wesley, Reading MA, Chapters 10 and 12 (1985).
3. A.T. Balaban, Constitutional and steric isomers of substituted cycloalkanes, Croat Chlem. Acta, **51** (1), 35-42 (1978).
4. K. Balasubramanian, Combinatorial enumeration of chemical isomers, Indian J. Chem., **16**B, 1094-1096 (1978).
5. S.Fujita, Subsymmetry-itemized enumeration of flexible cyclohexane derivatives, Bull. Chem. Soc. Japan, **67**, 2935-2948 (1994).
6. S. Fujita, Symmetry and combinatorial enumeration in chemistry, Springer-Verlag, Berlin, p. 50 (1991).
7. D. Michael P. Mingos and D.J. Wales, Introduction to cluster chemistry, Prentice-Hall Inc., London, pp. 62, 76 (1990).
8. K. Balasubramanian, Applications of combinatorics and graph theory to spectroscopy and quantum chemistry, Chem. Rev. **85** (6), pp. 602-603 (1985).
9. K. Balasubramanian, Computer-assisted enumeration of nmr signals, J. Magnetic resonance, **48**, 165-177 (1982); Applications of combinatorics and graph theory to spectroscopy and quantum chemistry, Chem. Rev., **85** (6), p. 604 (1985).

Chapter 3

The Enumeration Journey of Fujita : Subgroups, Cosets , Marks and Subductions of a Group

3.1 Introduction

The numerous applications of mathematics in chemistry (and physics) fall into two major areas: a) Applications involving differential and integral calculus. This type of mathematics is used to develop the language of chemical thermodynamics and design quantum-mechanical models of atomic and molecular structures, b) Applications which involve two types of discrete mathematics, viz., Graph Theory and Group Theory. Chemical applications of graph theory is now a well-defined topic usually called *chemical grap theory.*[1] This latter field considers among many chemical systems, in particular, benzenoid hydrocarbons[2] which seem to be a very versatile soil for chemical applications of graph theory and combinatorics. On the other hand group theory, even in its "chemical" form involves the uses of character tables to study such topics as Huckel molecular orbial model, ligand field theory and molecular vibrations[3]. These (chemical) applications, however, are very much oriented to the general field of physical chemistry and, with the exception of the (narrow field) of benzenoid hydrocarbons[2] do not attract the traditional organic chemist. During the last decade, however, *Shinsaku Fujita*[4] revived interest in the (already known but unpopular) theory of *marks of coset representations of a group*[5]. These numbers usually are listed in so-called **Mark Tables**. Fujita elegantly used these tables in the combinatorial enumeration of stereoisomers and to study chirality and other related stereochemical phenomena. Remarkably mark tables did not attract attention of neither chemists nor mathematicians for nearly eighty years since they were first published in 1911 ! By integrating together point-group theory and permutation group theory through marks of coset representations, Fujita considered and answered the following problem: Suppose a certain substitution pattern on a molecule whose parent skeleton (≡ parent molecular graph) belongs to a point-group **G**, where **G** includes the set of rotations, reflections and possibly rotoreflections which characterizes the parent skeleton. Suppose the set of representative subgroups of **G** has the following subgroups: $\mathbf{G}_1$, $\mathbf{G}_2$, ..., $\mathbf{G}_5$ ($\equiv \mathbf{G}$) . Then Fujita computes how many isomers possess $\mathbf{G}_1$ symmetry, how many belong to $\mathbf{G}_2$ symmetry, ... , $\mathbf{G}_S$ symmetry. This achievement is to be compared with the result of Pólya's theorem. **The latter can only generate the isomers but does not classify them according to their subgroups.** The count of stereoisomers and their classification according to their point-groups is certainly an important result to the organic chemists. However, Fujita's theory uses quite abstract concepts (such as **cost representations,** their **marks** and **group subduction tables).**

Such terms are not likely to assume popular or attractive vocabulary among traditional organic chemists. Indeed a paradox can be seen in the following way:

a) The work of Fujita requires people who are willing to think mathematically[6] and be able to "open" an abstract concept into a real observable;
b) The **results** of the work of Fujita serve mostly organic chemists;
c) The majority of organic chemists prefer to stay away from abstraction and may be repelled by mathematics.

Whence a) , b) and c) lead to a paradox !

In other words this important work will be read and understood by theoretical chemists who may *not* be interested in organic chemistry!

Another barrier, which is not related to the mathematical nature of coset representation theory, is the "stereochemical" barrier. This is "educational" in nature and involves how to represent three-dimensional molecules, such as adamantane, cubane and even molecules as simple as allene on a (2-dimensional) paper plane. This second barrier is related to how one envisages a given symmetry operation, its permutation representation and whence the resulting S_i^j notations. The latter, of course, is quite essential to the structure of the cycle index and other related functions.

The situation is sketched in FIG 35 in which an (organic) chemist is trying to reach the "last station" in the enumeration journey of Fujita which is the **Isomer-Count (IC) matrix**, what we call: **the heavens of Fujita !.**

Figure 35 (see page 185)

Naturally, at this point, we do not expect a reader to *understand* what is going on, but we only wanted to prepare the reader for the kind of ground he will soon be landing on!

The main goal of this book is to solve this paradox by casting the theory put forward by Fujita in the form of a simple graphical model which can be comprehensed with only a minimum knowledge of symmetry. Namely, being able **only** to rotate and reflect planar graphs !. This model, however, will be described in the next chapter. The strategy of the present chapter is to consider the steps involved in the enumeration scheme of Fujita, then define the terms involved and consider as many examples as possible to familiarize the reader with this (new) language. We will do the above steps with the minimum possible of mathematics. After a reader had solved an actual example of enumeration, he should be ready to learn our model described in chapter 4.

3.2. The Enumeration Scheme of Fujita

The following steps are essential to generate all isomers which result from a given skeleton *and* classify them according to their subgroup symmetry. These steps are graphed and their inter-relations clarified in FIG 36. These steps will be denoted here as the **enumeration scheme (or in fact: " journey") of Fujita.**

Figure 36 (see page 186)

We now mention these steps, then define and exemplify the terms involved: [All terms are defined in ref. 4].

1) Find the point-group of the parent skeleton (i.e., the molecular graph) to be substituted and its **multiplication table**, C.f. Table 1, Chapter 1.

2) Determine all subgroups of the point-group and arrange them in a non-descending order, viz., $\mathbf{G}_1, \mathbf{G}_2, \ldots, \mathbf{G}_S$; where $|\mathbf{G}_1| \leq |\mathbf{G}_2| \leq \ldots \leq |\mathbf{G}_S|$ where $\mathbf{G}_S = \mathbf{G}$ = the point group of the parent skeleton and $\mathbf{G}_1 \equiv \mathbf{C}_1 = \{I\}$. In general $|\mathbf{G}_i|$ is the order of the ith subgroup = the number of elements in it. If two (or more) subgroups are **conjugate**[7] only one is selected. The resulting subgroup sequence is called representative **sequence of subgroups. SSG.**

3) Step 2) is essential to find all **coset representations,** CR,of the group under consideration.

4) Calculatge the **marks** associated with each CR for each subgroup. I.e., each CR generates a row of marks. When the mark rows are assembled in the form of a table whose columns are headed by the individual subgroups the so-called **Mark Table** results.

5) Determine the CR which controls the **orbit,** Δ, of substitution, i.e., the set of vertices which undergoes substitution.

6) **Subduce** the CR reached in step 5 by all subgroups of the parent point-group.

7) Associate each subduction of the controlling CR with what Fujita calls, **Unit Subduced Cycle Index**, USCI. We will have as many of such cycle indices as there are subgroups.

8) The resulting USCI's are then incorporated into a Pólya-type substitution, viz.,

$$S^{n}_{m} \rightarrow (x^{m} + y^{m} + z^{m} + \ldots)^{n} \qquad \ldots (109)$$

where, as usual, x, y, z, ... , are assigned "weights" of substituents. Eqn. (109) generates a polynomial in xyz ... , the various powers of which form the so-called **Fixed Point Matrix**, FPM. The sums of columns of the FPM generate a **Fixed Point Vector**, FPV, which may be generalized as :

$$\text{FPV} = \mathbf{A} = (A_{\mathbf{G}1}\ A_{\mathbf{G}2} \ldots A_{\mathbf{G}s}) \qquad \ldots (110)$$

Where A_{G_i} is the number of isomers which remain fixed (i.e. invariant) under the symmetry operations of $\mathbf{G}_i$, where $1 \leq I \leq s$.

9) The final step in this enumeration journey is to use the mark table reached in step 4 to construct the **Isomer Count Matrix**, ICM. The sum of the rows of this matrix is an Isomer Count Vector, ICV, given by :

$$\mathbf{A} = (A_{G_1}\ A_{G_2} \ldots A_{G_s}) = \rho_1(m_{11}) + \rho_2(m_{21} + m_{22}) + \ldots + \rho_i(m_{i1} + m_{i2} + .. + m_{ii}) + \rho_s(1\ 1\ 1 \ldots 1) \qquad \ldots (111)$$

where m_{ij} is the number of cosets which remain fixed in the ith coset representation under the effect of symmetry operations of the jth subgroup of the parent point-group. The number of 1's in eqn. (111) = | SSG | = the number of representative subgroups of **G** , c.f. step 2.

Eqn. (111) says that there are ρ_1 isomers which possess $\mathbf{G}_1$ symmetry, ρ isomers which possess $\mathbf{G}_2$ symmetry and so on. Then an alternative form of eqn. (111) is :

$$\mathbf{A}\,\mathbf{M}^{-1} = (\rho_2\ \rho_1 \ldots \rho_s) \equiv \Lambda \qquad \ldots (112)$$

Where $\mathbf{M}^{-1}$ is the inverse of the mark table. (Recall that mij's of eqn. (111) are the elements of the mark table !). Here a reader may slightly benefit, at this point, by looking through FIG 36. Now we shall go into the (difficult) task of explaining and exemplifying these (new) terms.

3.2.1. Subgroups of a Group

The set of symmetry elements { g_1 , g_2 , … , $g_{|G|}$ } which defines a point-group **G** (of order |**G**|) can often be decomposed into subsets, each of which satisfy the axioms (properties) of a group mentioned in section 1.2 of Chapter 1.

Example 31 A simple graph but not so simple point-group

There are 8 subgroup of the $\mathbf{D}_4$ point-group shown and modeled in FIG 1.

This can easily be verified by using its multiplication table, table 1, [C.f. Chapter 1].

There is no systematic way in which we can find all subgroups of a given group. However we can always find a "chemical representation " of a given subgroup. As an illustration we consider the general system shown below:

$$\begin{array}{ccc} X & — & X \\ | & & | \\ X & — & X \end{array} \quad ; \; X = \{ C, N \} \qquad \ldots (113)$$

One can easily use theorem of Pólya to derive all derivatives which have this general formula. These are shown in FIG 37. We observe that each derivative remains fixed under a subset of symmetry elements (derived from $\mathbf{D}_4$) which defines a subgroup. This way we defined four of the subgroups of this point group.

Figure 37 (see page 187)

3.2.2 Coset representation, CR, of a group[4,7]

Normally one defines a group by indicating the set of elements that are contained in that group. This is shown in FIG 1,2 and 3 for several point-groups. However, there is another representation in which the full group is generated through one of its subgroups For example, the subgroup $\mathbf{C}_4^2 = \{ I, C_4^2 \}$ may be used to generate all the elements in $\mathbf{D}_4$ by using the multiplication table, Table 1, we can write :

$$\begin{aligned} \mathbf{C}_4^2 &= \{ I, C_4^2 \} \\ \mathbf{C}_4^2 C_4 &= \{ IC_4, C_4^2 C_4 \} = \{ C_4, C_4^3 \} \\ \mathbf{C}_4^2 C_2 &= \{ IC_2, C_4^2 C_2 \} = \{ C_2, C_2' \} \\ \mathbf{C}_4^2 C_{2(d)} &= \{ IC_{2(d)}, C_4^2 C_{2(d)} \} = \{ C_{2(d)}, C_{2(d)}' \} \end{aligned} \qquad \ldots (114)$$

The elements of the right-hand side of eqn. (114) defines **D**$_4$. The symbol **D**$_4$ $(/\mathbf{C}_4^2)$ is called a (right) coset representation of **D**$_4$ by **C**$_4$ and we can write

$$\mathbf{D}_4\,(/\mathbf{C}_4^2) = \mathbf{C}_4^2 + \mathbf{C}_4^2\,C_4 + \mathbf{C}_4^2\,C_2 + \mathbf{C}_4^2\,C_{2(d)} \qquad \ldots (115)$$

3.2.3 Permutation representation, PR, of a group

Let us label the components of the CR : **D**$_4$ $(/\mathbf{C}_4^2)$ by 1,2,3 and 4 and multiply each component by all the elements of **D**$_4$. The result is outlined below in Table 3:

Table 3

Permutation representation of **D**$_4$ $(/\mathbf{C}_4^2)$

	$\{I, C_4^2\}$	$\{C_4, C_4^3\}$	$\{C_2, C_2'\}$	$\{C_{2(d)}, C_{2(d)}'\}$	
CR : $\mathbf{D}_4(/\mathbf{C}_4^2)$ =	$\mathbf{C}_4^2$ +	$\mathbf{C}_4^2 C_4$ +	$\mathbf{C}_4^2 C_2$ +	$\mathbf{C}_4^2 C_{2(d)}$	PR
	1	2	3	4	
I	1	2	3	4	(1)(2)(3)(4)
C_4	2	1	4	3	(12) (34)
C_4^2	1	2	3	4	(1)(2)(3)(4)
C_4^3	2	1	4	3	(12) (34)
C_2	3	4	1	2	(13) (24)
C_2'	3	4	1	2	(13) (24)
$C_{2(d)}$	4	3	2	1	(14) (23)
$C_{2(d)}'$	4	3	2	1	(14) (23)

The column of the PR`s lists the permutations relative to the first row. For example the fourth row is derived as follows : (Use multiplication table, Table 1).

$$\mathbf{C}_4^2 C_4^3 = \{I, C_4^2\}\, C_4^3 = \{C_4^3, C_4\} \equiv \mathbf{C}_4^2 C_4 \equiv 2$$

$$\mathbf{C}_4^2 C_4 C_4^3 = \{I, C_4^2\} \equiv \mathbf{C}_4^2 \equiv 1$$

$$\mathbf{C}_4^2 C_2 C_4^3 = \{C'_{2(d)}, C_{2(d)}\} = \mathbf{C}_4^2 C_{2(d)} \equiv 4$$

$$\mathbf{C}_4^2 C_{2(d)} C_4^3 = \{C_2 C'_2\} = \mathbf{C}_4^2 C_2 \equiv 3$$

The resuting permutation, then, is

$$\begin{pmatrix} 1 & 2 & 3 & 4 \\ 2 & 1 & 4 & 3 \end{pmatrix} \equiv (12)(34) \qquad \dots (116)$$

Permutation representation of a CR, $\mathbf{G}(\ /\mathbf{G}_i)$ relative to a given subgroup $\mathbf{G}_j$ is associated with a mark m_{ij} as we explain in the next section.

3.2.4 The marks of a given CR : Table of Marks [4]

A mark m_{ij}, is best defined by the following steps:

1. Find the CR $\mathbf{G}(\ /\mathbf{G}_i)$ and the corresponding PR using all elements of **G**.
2. Mark the elements of the subgroups $\mathbf{G}_j$ by √
3. Count the number of fixed cosets under $\mathbf{G}_j$ in the rows marked by √.

The resulting number is m_{ij}.

When these steps are repeated for all values of i and j (observe that $1 \le i, j \le s$), i.e. for all CR's under the symmetry operations of all subgroups the table of marks (or mark table) of that point-group results. In the mark table the subgroups are listed in a non-descending order [C.f. section 3.2].

Example: 32

Find the row of marks which correspond to $\mathbf{D}_4(\ /\mathbf{C}_4^2)$.

Firstly we find the sequence of subgroups, SSG, of $\mathbf{D}_4$ listed in a nondescending way as follows:

$\mathbf{G}_1$	$\mathbf{G}_2$	$\mathbf{G}_3$	$\mathbf{G}_4$	$\mathbf{G}_5$	$\mathbf{G}_6$	$\mathbf{G}_7$	$\mathbf{G}_8$
$\mathbf{C}_1$	$\mathbf{C}_4$	$\mathbf{C}_2$	$\mathbf{C}_S$	$\mathbf{C}_4$	$\mathbf{C}_{2v}$	$\mathbf{D}_2$	$\mathbf{D}_4$

... (117)

Now we must find the subgroups in the subgroup $\mathbf{C}_4^2 = \{I, C_4^2\}$. There are two subgroups, viz., $\mathbf{C}_1 = \{I\}$ and $\mathbf{C}_4^2$. Now we re-list the PR's calculated for $\mathbf{D}_4(/\mathbf{C}_4^2)$ from Table 3 and mark the rows which correspond to subgroups of $\mathbf{C}_4^2$. We have :

		$G_1 = \{I\}$	$G_2 = \mathbf{C}_4^2$
I	(1)(2)(3)(4)	√	√
C_4	(12) (34)		
C_4^2	(1)(2)(3)(4)		√
C_4^3	(12) (34)		
C_2	(13) (24)		
C_2'	(13) (24)		
$C_{2(d)}$	(14) (23)		
$C_{2(d)}'$	(14) (23)		
		$m_{21} = 4$	$m_{22} = 4$

... (118)

The above search for marks may be outlined in the form of the following row:

	$\mathbf{C}_1$	$\mathbf{C}_4^2$	$\mathbf{C}_2$	$\mathbf{C}_s$	$\mathbf{C}_4$	$\mathbf{C}_{2v}$	$\mathbf{D}_2$	$\mathbf{D}_4$	
$\mathbf{D}_4(/\mathbf{C}_4^2)$	(4	4	0	0	0	0	0	0)	... (119)

We observe that because $\mathbf{C}_4^2$ contains only two subgroups (labeled in our nondescending sequence as G_1 and G_2) only m_{21} and m_{22} have nonvanishing terms. Indeed m_{ij} for $j > i$ are always zero.

Now we detail the generation of the mark table of $\mathbf{D_4}$.

1. $\mathbf{D_4(/C_1)} = \mathbf{C_1} + \mathbf{C_1}\, C_4 + \mathbf{C_1} C_4^2 + \mathbf{C_1} C_4^3 + \mathbf{C_1}\, C_2 + \mathbf{C_1}\, C_2' + \mathbf{C_1}\, C_{2(d)} + C_1\, C_{2(d)}'$

	1	2	3	4	5	6	7	8
I	1	2	3	4	5	6	7	8
→	(1)(2)(3)(4)(5)(6)(7)(8)							√
C_4	2	3	4	1	7	8	6	5
→	(1 2 3 4)(5 7 6 8)							
C_4^2	3	4	1	2	6	5	8	7
→	(13) (24) (56) (78)							
C_4^3	4	1	2	3	8	7	5	6
→	(1432) (5867)							
C_2	5	8	6	7	1	3	4	2
→	(15) (28) (36) (47)							
C_2'	6	7	5	8	3	1	2	4
→	(16) (27) (35) (48)							
$C_{2(d)}$	7	5	8	6	2	4	1	3
→	(17) (25) (38) (46)							
$C_{2(d)}'$	8	6	7	5	4	2	3	1
→	(18) (26) (37) (45)							

Mark row:

	$\mathbf{C_1}$	$\mathbf{C_4^2}$	$\mathbf{C_2}$	$\mathbf{C_s}$	$\mathbf{C_4}$	$\mathbf{C_{2v}}$	$\mathbf{D_2}$	$\mathbf{D_4}$	
$\mathbf{D_4(/C_1)}$	(8	0	0	0	0	0	0	0)	... (120)

2. $\mathbf{D_4}\,(/\ \mathbf{C_4^2})$

This has already been computed, Table 3, eqns. (118) , (119)

3. $\mathbf{D_4(/C_2)} = \mathbf{C_2} + \mathbf{C_2 C_4} + \mathbf{C_2} C_4^2 + \mathbf{C_2} C_4^3$ PR

	1	2	3	4		C_1	C_2
I	1	2	3	4	(1)(2)(3)(4)	√	√
C_4	2	3	4	1	(1234)		
C_4^2	3	4	1	2	(13) (24)		
C_4^3	4	1	2	3	(1432)		
C_2	1	4	3	2	(1) (3) (24)		√
C_2'	3	2	1	4	(2) (4) (13)		
$C_{2(d)}$	2	1	4	3	(12) (34)		
$C_{2(d)}'$	4	3	2	1	(14) (23)		

Mark row:

	C_1	C_4^2	C_2	C_s	C_4	C_{2v}	D_2	D_4	
$D_4(/C_2)$	(4	0	2	0	0	0	0	0)	... (121)

4. $\mathbf{D_4(/C_s)} = \mathbf{C_s} + \mathbf{C_s C_4} + \mathbf{C_s} C_4^2 + \mathbf{C_s} C_4^3$ PR

	1	2	3	4		C_1	C_s
I	1	2	3	4	(1)(2)(3)(4)	√	√
C_4	2	3	4	1	(1234)		
C_4^2	3	4	1	2	(13) (24)		
C_4^3	4	1	2	3	(1432)		
C_2	4	3	2	1	(14) (23)		
C_2'	2	1	4	3	(12) (34)		
$C_{2(d)}$	1	4	3	2	(1) (3) (24)		√
$C_{2(d)}'$	3	2	1	4	(2) (4) (13)		

Mark row:

	C_1	C_4^2	C_2	C_s	C_4	C_{2v}	D_2	D_4	
$D_4(/C_s)$	(4	0	0	2	0	0	0	0)	... (122)

5. $\mathbf{D_4(/\,C_4)} = \mathbf{C_4 + C_4\,C_2}$

	1	2	PR	C_1	C_4^2	C_4
I	1	2	(1)(2)	√	√	√
C_4	1	2	(1)(2)			√
C_4^2	1	2	(1)(2)		√	√
C_4^3	1	2	(1)(2)			√
C_2	2	1	(12)			
C_2'	2	1	(12)			
$C_{2(d)}$	2	1	(12)			
$C'_{2(d)}$	2	1	(12)			

Mark row:

	C_1	C_4^2	C_2	C_s	C_4	C_{2v}	D_2	D_4	
$D_4(/C_4)$	(2	2	0	0	2	0	0	0)	… (123)

6. $\mathbf{D_4(/\,C_{2v})} = \mathbf{C_{2v} + C_{2v}\,C_4}$

	1	2	PR	C_1	C_4^2	C_2	C_s	C_{2v}
I	1	2	(1)(2)	√	√	√	√	√
C_4	2	1	(12)					
C_4^2	1	2	(1)(2)		√			√
C_4^3	2	1	(12)					
C_2	2	1	(12)			√		
C_2'	2	1	(12)					
$C_{2(d)}$	1	2	(1)(2)				√	√
$C'_{2(d)}$	1	2	(1)(2)					√

Mark row:

	C_1	C_4^2	C_2	C_s	C_4	C_{2v}	D_2	D_4	
$D_4(/C_{2v})$	(2	2	0	2	0	2	0	0)	… (124)

7. $\mathbf{D_4(/\,C_2)}$ = $\mathbf{D_2}$ + $\mathbf{D_2\,C_4}$

	1	2	PR	C_1	C_4^2	C_2	D_2
I	1	2	(1)(2)	√	√	√	√
C_4	2	1	(12)				
C_4^2	1	2	(1)(2)		√		√
C_4^3	2	1	(12)				
C_2	1	2	(1)(2)			√	√
C_2'	1	2	(1)(2)				√
$C_{2(d)}$	2	1	(12)				
$C_{2(d)}'$	2	1	(12)				

Mark row:

	C_1	C_4^2	C_2	C_S	C_4	C_{2v}	D_2	D_4	
$D_4(/D_2)$	(2	2	2	0	0	0	2	0)	… (125)

8. $\mathbf{D_4(/\,D_4)}$ = $\mathbf{D_4}$

	1	PR	C_1	C_4^2	C_2	C_S	C_5	C_{2v}	D_2	D_4
I	1	(1)	√	√	√	√	√	√	√	√
C_4	1	(1)					√			
C_4^2	1	(1)		√			√	√	√	√
C_4^3	1	(1)					√			√
C_2	1	(1)			√				√	√
C_2'	1	(1)							√	√
$C_{2(d)}$	2	(1)				√		√		√
$C_{2(d)}'$	1	(1)						√		

Mark row:

	C_1	C_4^2	C_2	C_S	C_4	C_{2v}	D_2	D_4	
$D_4(/C_4)$	(1	1	1	1	1	1	1	1)	… (126)

The collection of the row of marks [eqns (119) – (126) leads to the Table of Marks (or simply, Mak Table) of this point-group. This is given below in Table 4 (Mark tables of several point-groups are given in the appendix of ref 4).

Table 4

Mark table of $\mathbf{D_4}$

	C_1	C_4^2	C_2	C_S	C_4	C_{2v}	D_2	C_4
$\mathbf{D_4(/C_1)}$	8	0	0	0	0	0	0	0
$\mathbf{D_4(/C_4^2)}$	4	4	0	0	0	0	0	0
$\mathbf{D_4(/C_2)}$	4	0	2	0	0	0	0	0
$\mathbf{D_4(/C_S)}$	4	0	0	2	0	0	0	0
$\mathbf{D_4(/C_4)}$	2	2	0	0	2	0	0	0
$\mathbf{D_4(/C_{2v})}$	2	2	0	2	0	2	0	0
$\mathbf{D_4(/D_2)}$	2	2	2	0	0	0	2	0
$\mathbf{D_4(/D_4)}$	1	1	1	1	1	1	1	1

Mark tables of groups have the following apparent properties :

1) m_{11} = number of elements in **G** , i.e. $m_{11} = |\mathbf{G}|$. This is because the particular CR : $\mathbf{G(/C_1)}$ simply lists all elements in **G** in the particular PR which corresponds to the identity element, C.f. derivation of equation (120).
2) The last row is always composed of ones. This is a direct result of the multiplication properties of the elements of a group.
3) All m_{ij}'s are zero for all $j > i$.
4) The number of columns = the number of rows and as a consequence of property 3, the upper triangle above the main diagonal is composed of zeros.

3.2.5 The orbit which controls substitution in the parent graph[4,8]

The concept of an orbit, Δ , has been introduced in section 2.5.3.2 when we studied ^{13}C nmr. We know from FIG 34 that there are three orbits in the H-suppressed naphthalene graph, while anthracene has 4 orbits, pyrene has five orbits and finally coronene has 3 orbits. *Fujita demonstrated that each orbit is associated with a particular CR* [4,8].

It is crucial to specify the so-called orbit which controls substitution in an enumeration scheme. As an illustration we consider the square which belongs to $\mathbf{D_4}$ now we apply all elements in this point-group to the four vertices of the square and count the number of fixed ones under the symmetry operations of all subgroups in $\mathbf{D_4}$. This is shown in Table 5.

Table 5

The effect of symmetry operations on the four vertices of the square. The vertices are labeled 1,2,3,4 in sequence and the symmetry elements are shown in FIG 1. Mark row is concluded in eqn. (127).

		$\mathbf{C_1}$	$\mathbf{C_4^2}$	$\mathbf{C_2}$	$\mathbf{C_S}$	$\mathbf{C_4}$	$\mathbf{C_{2v}}$	$\mathbf{D_2}$	$\mathbf{D_4}$
I	(1)(2)(3)(4)	√	√	√	√	√	√	√	√
C_4	(1432)					√			√
C_4^2	(13)(24)		√			√	√	√	√
C_4^3	(1234)					√			√
C_2	(12)(34)			√				√	√
C_2'	(14)(23)							√	√
$C_{2(d)}$	(2)(4)(13)				√		√		√
$C_{2(d)}'$	(1)(3)(24)						√		√

	$\mathbf{C_1}$	$\mathbf{C_4^2}$	$\mathbf{C_2}$	$\mathbf{C_S}$	$\mathbf{C_4}$	$\mathbf{C_{2v}}$	$\mathbf{D_2}$	$\mathbf{D_4}$	
Row of marks: (	4	0	0	2	0	0	0	0)	... (127)

Now we compare the row of marks computed in eqn. (127) with the mark table of this point-group, Table 4, and we conclude that the orbit under consideration (Δ = the set of 4 vertices of the square) corresponds to the CR $\mathbf{D_4}$ (/$\mathbf{C_S}$). We, then, say that the orbit of substitution in the square is controled by $\mathbf{D_4}$ (/$\mathbf{C_S}$).

So far we have defined subgroups, coset representations, and marks of a group. Also, the important concept of the orbit which controls substitution in a given parent graph. The next step now, is to learn how to *subduce*[8,9] the orbit of substitution by all the subgroups of the parent group. *This requires mark tables of all the subgroups of the point-group under consideration.* Therefore before we consider the **group-subduction formula of Fujita** we give the mark tables of the subgroups of $\mathbf{D}_4$.

Table 6

Mark table of $\mathbf{C}_4^2$

	$\mathbf{C}_1$	$\mathbf{C}_4^2$
$\mathbf{C}_4^2$ (/$\mathbf{C}_1$)	2	0
$\mathbf{C}_4^2$ (/$\mathbf{C}_4^2$)	1	1

Table 7

Mark table of $\mathbf{C}_2$

	$\mathbf{C}_1$	$\mathbf{C}_2$
$\mathbf{C}_2$ (/$\mathbf{C}_1$)	2	0
$\mathbf{C}_2$ (/$\mathbf{C}_2$)	1	1

Table 8

Mark table of $\mathbf{C}_S$

	$\mathbf{C}_1$	$\mathbf{C}_S$
$\mathbf{C}_5$ (/$\mathbf{C}_1$)	2	0
$\mathbf{C}_5$ (/$\mathbf{C}_S$)	1	1

Table 9

Mark table of $\mathbf{C}_4$

	$\mathbf{C}_1$	$\mathbf{C}_4^2$	$\mathbf{C}_4$
$\mathbf{C}_2$ (/$\mathbf{C}_1$)	4	0	0
$\mathbf{C}_4$ (/$\mathbf{C}_4^2$)	2	2	0
$\mathbf{C}_4$ (/$\mathbf{C}_4$)	1	1	1

Table 10

Mark table of $\mathbf{C_{2v}}$

	C_1	C_4^2	C_s	C_s'	C_{2v}
$C_{2v}(/C_1)$	4	0	0	0	0
$C_{2v}(/C_4^2)$	2	2	0	0	0
$C_{2v}(/C_s)$	2	0	2	0	0
$C_{2v}(/C_s')$	2	0	0	2	0
$C_{2v}(/C_{2v})$	1	1	1	1	1

Table 11

Mark table of $\mathbf{D_2}$

	C_1	C_4^2	C_s	C_s'	D_2
$D_2(/C_1)$	4	0	0	0	0
$D_2(/C_4^2)$	2	2	0	0	0
$D_2(/C_2)$	2	0	2	0	0
$D_2(/C_s')$	2	0	0	2	0
$D_2(/D_2)$	1	1	1	1	1

The mark table is already given in Table 4.

3.2.6 The subduction formula of Fujita: Unit Subduced Cycle Index, USCI[4,8,9]

Fujita derives what he called USCI's and uses them in Pólya-type substitution to find the fixed-point matrix. These cycle indices are derived by first subducing the controlling orbit by all subgroups of the parent group. For the case at hand this orbit turns out to be $\mathbf{D_4}(/\mathbf{C_S})$, c.f. eqn. (127). To find $\mathbf{D_4}(/\mathbf{C_S}) \downarrow \mathbf{G_i}$, i.e., the subduction of $\mathbf{D_4}(/\mathbf{C_S})$ by $\mathbf{G_i}$, one of the subgroups of $\mathbf{D_4}$, Fujita considers the set of permutations of $\mathbf{D_4}(/\mathbf{C_S})$ which remains after eliminating all permutations except those which correspond to $\mathbf{G_i}$. For example to find $\mathbf{D_4}(/\mathbf{C_S}) \downarrow \mathbf{G_{2v}}$ we look at Table 5 and copy only those permutations which correspond to the elements of $\mathbf{C_{2v}}$ and compute the row of marks generated under the symmetry operations of all subgroups of $\mathbf{C_{2v}}$. This is done below in Table 12.

Table 12

Subduction of $\mathbf{D}_4$ ($/\mathbf{C}_S$) by $\mathbf{C}_{2v}$. See Table 5.

		$\mathbf{C}_1$	$\mathbf{C}_4^2$	$\mathbf{C}_S$	$\mathbf{C}'_s$	$\mathbf{C}_{2v}$
I	(1)(2)(3)(4)	√	√	√	√	√
C_4^2	(13)(24)		√			√
$C_{2(d)}$	(1)(3)(24)			√		√
$C'_{2(d)}$	(2)(4)(13)				√	√

Row of marks: (4 0 2 2 0) ... (128)

Now we look at Table 10, the mark table of the $\mathbf{C}_{2v}$ point group, and discover that if we add together the mark rows of $\mathbf{C}_{2v}(/\mathbf{C}_S) + \mathbf{C}_{2v}(/\mathbf{C}'_s)$ we obtain the row of marks generated in eqn (128), i.e.,

	$\mathbf{C}_1$	$\mathbf{C}_4^2$	$\mathbf{C}_S$	$\mathbf{C}'_s$	$\mathbf{C}_{2v}$
$\mathbf{C}_{2v}(/\mathbf{C}_S)$	(2	0	2	0	0)
$\mathbf{C}_{2v}(/\mathbf{C}'_s)$	(2	0	0	2	0)
	(4	0	2	2	0) ... (129)

And whence we may write:

$$\mathbf{D}_4\,(/\mathbf{C}_S) \downarrow \mathbf{C}_{2v} = \mathbf{C}_{2v}\,(/\mathbf{C}_S) + \mathbf{C}_{2v}(/\mathbf{C}'_s) \quad \ldots (130)$$

A more general way of expressing this subduction operation is the following:

$$\mathbf{G}\ (/\mathbf{G}_i) \downarrow \mathbf{G}_j = \beta_{jk}\,\mathbf{G}_j\,(/\mathbf{G}_k) + \beta_{j\ell}\,\mathbf{G}_j\,(/\mathbf{G}_\ell) \quad \ldots (131)$$

Where $\mathbf{G}_i$, $\mathbf{G}_j$, $\mathbf{G}_k$, $\mathbf{G}_\ell$, ... are all subgroups of the present group $\mathbf{G}$, while $\mathbf{G}_k$, $\mathbf{G}_\ell$, ... are subgroups of the subducing subgroup. The parameters β_{jk}, ... are multiplicity factors.

The formula of Fujita

Fujita rigorously concludes that the general form of eqn. (131) corresponds to the following USCI:

$$S_{d_{jk}}^{\beta_{jk}} \quad S_{d_{j\ell}}^{\beta_{j\ell}} \qquad \text{... (132)}$$

where $d_{jk} = |G_j| \,/\, |G_k|$;

$$d_{j\ell} = |G_j| \,/\, |G_\ell| \qquad \text{... (133)}$$

in which $|G_j|$ = order of G_j (i.e. number of elements in it) and so on. Then for eqn. (130) we have :

$|C_{2v}| \,/\, |C_s| = 4/2 = 2$;

$|C_{2v}| \,/\, |C'_s| = 4/2 = 2$;

and since the β's (multiplicity factors) are unities, we can write that subduction of this CR by C_{2v} generates a USCI = S_2^2 , viz.,

$$D_4\,(/C_s) \downarrow C_{2v} \sim S_2\,S_2 = S_2^2 \qquad \text{... (134)}$$

Throughout the following we derive all USCI's for all other subductions of this CR. In all cases the permutations are copied from Table 5 and we use mark tables of the relevant subgroups, Tables 6-11 as well as the mark table of the parent group, Table 4.

1) $D_4\,(/C_s) \downarrow C_1 \rightarrow$ [(1)(2)(3)(4)] $\rightarrow 4\,C_1\,(/C_1) \rightarrow S_1^4$ ... (135)

2) $D_4\,(/C_s) \downarrow C_4^2 \rightarrow$ [(1)(2)(3)(4) / (13) (24)] $\rightarrow (4 \quad 0) \rightarrow$

$2\,C_4^2\,(/C_1) \rightarrow S_2^2$ ... (136)

3) $D_4\,(/C_s) \downarrow C_2 \rightarrow$ [(1)(2)(3)(4) / (14) (23)] $(4 \quad 0) \rightarrow$

$2\;C_2\;(/C_1) \rightarrow S_2^2$ ... (137)

4) $\mathbf{D_4}$ (/$\mathbf{C_s}$) ↓ $\mathbf{C_s}$ →

(1)(2)(3)(4)
(1) (3)(24)

→ (4 2) →

$\mathbf{C_s}$ (/$\mathbf{C_1}$) + 2 $\mathbf{C_s}$(/$\mathbf{C_s}$) → $S_2 S_1^2 \equiv S_1^2 S_2$ ... (138)

5) $\mathbf{D_4}$ (/$\mathbf{C_s}$) ↓ $\mathbf{C_4}$ →

(1)(2)(3)(4)
(1234)
(13)(24)
(1432)

→ (4 0 0) →

$\mathbf{C_4}$ (/$\mathbf{C_1}$) → S_4 ... (139)

6) $\mathbf{D_4}$ (/$\mathbf{C_s}$) ↓ $\mathbf{C_{2v}}$ → → → S_2^2 (already derived) ... (140)

7) $\mathbf{D_4}$ (/$\mathbf{C_s}$) ↓ $\mathbf{D_2}$ →

(1)(2)(3)(4)
(13)(24)
(14)(23)
(12)(34)

→ (4 0 0 0 0) →

$\mathbf{D_2}$ (/$\mathbf{C_1}$) → S_4 ... (141)

8) $\mathbf{D_4}$ (/Cs) ↓ $\mathbf{D_2}$ →

(1)(2)(3)(4)
(1234)
(13)(24)
(1432)
(14)(23)
(12)(34)
(1)(3)(24)
(2)(4)(13)

→ (4 0 0 2 0 0 0 0) →

→ $\mathbf{D_4}$ (/$\mathbf{C_5}$) → S_4 ... (142)

* * *

Now, that we have defined, and exemplified these concepts (marks of a group, orbit of substitution, subduction of a coset representation and unit subduced cycle index) we are prepared to go through one of the enumeration journies of Fujita through the following illustration:

Example 33 Generation of heterocycles, oncemore!

Find all derivatives consistent with the following general formula

$$\begin{array}{ccc} X & \text{—} & X \\ | & & | \\ X & \text{—} & X \end{array} \quad ; \; X = \{ C, N \} \qquad \ldots (143)$$

and assign them to their proper subgroups. I.e., compute their Isomer-Count (I.C) matrix.

It is better at this stage that a reader looks carefully into FIG 36 and try to understand how the steps in the enumeration scheme of Fujita are inter-related. Clearly this is a $\mathbf{D_4}$ problem with which we are familiar by now !. The mark table of the point group is given in Table 4, the orbit which controls substitution is a $D_4(/C_S)$ type, c.f. eqn. (127). All the necessary subductions and the resulting USCI's are generated in eqns. (135)-(142). All that is left is to find the fixed-point matrix (FPM) and isomer-count matrix (ICM). The former requires one to apply Pólya substitution into the calculated USCI's. This is done below : (assuming w(C) = 1 ; w(N) = x).

1) $\mathbf{C_1} \rightarrow s_1^4 \rightarrow (1+x)^4 = 1 + 4x + 6x^2 + 4x^3 + x^4$ ... (144)

2) $\mathbf{C_4^2} \rightarrow s_2^2 \rightarrow (1+x^2)^2 = 1 + 2x^2 + x^4$ ... (145)

3) $\mathbf{C_2} \rightarrow s_2^2 \rightarrow (1+x^2)^2 = 1 + 2x^2 + x^4$ ... (146)

4) $\mathbf{C_S} \rightarrow s_1^2 s_2 \rightarrow (1+x)^2(1+x^2) = 1 + 2x + 2x^2 + 2x^3 + x^4$ ... (147)

5) $\mathbf{C_4} \rightarrow s_4^1 \rightarrow (1+x^4)^1 = 1 + x^4$ ... (148)

6) $\mathbf{C_{2v}} \rightarrow s_2^2 \rightarrow (1+x^2)^2 = 1 + 2x^2 + x^4$ ... (149)

7) $\mathbf{D_2} \rightarrow s_4^1 \rightarrow (1+x^4)^1 = 1 + x^4$ ... (150)

8) $\mathbf{D_4} \rightarrow s_4^1 \rightarrow (1+x^4)^1 = 1 + x^4$ ... (151)

The above polynomials are collected together to form the FPM given below:

Table 13

Fixed-Point Matrix (FPM) which corresponds to eqns. (144) – (151).

	C_1	C_4^2	C_2	C_s	C_4	C_{2v}	D_2	D_4
1	1	1	1	1	1	1	1	1
x	4	0	0	2	0	0	0	0
x^2	6	2	2	2	0	2	0	0
x^3	4	0	0	2	0	0	0	0
x^4	1	1	1	1	1	1	1	1

The sum of the coefficients lead to the following Fixed-Point Vector, FPV:

$$\begin{matrix} C_1 & C_4^2 & C_2 & C_s & C_4 & C_{2v} & D_2 & D_4 \\ (16 & 4 & 4 & 8 & 2 & 4 & 2 & 2) \end{matrix} \quad \ldots (152)$$

Eqn. (152) is one (open) form of eqn. (110), e.g., for our particular problem $A_{C_2} = 4$ and so on.

To construct the isomer count matrix, ICM, we extract from FPM the appropriate rows or sum of rows (multiplied with proper multiplication factors) from the table of marks of the $\mathbf{D_4}$ point-group, Table 3, which correspond to this FPM. For example, the first row in Table 13 corresponds to the last row of the mark table, then we write a 1 under $\mathbf{D_4}$ in the isomer-count matrix. (because the last row in the mark table corresponds to $\mathbf{D_4}(/\mathbf{D_4})$). Analogously the second row in our FPM corresponds to $\mathbf{D_4}/\mathbf{C_s}$) in the mark table and whence we write a 1 under $\mathbf{C_s}$ in the ICM. Sometimes one needs sum of more than one row : the vector fo x^2 , i.e. the third row in the FPM is the sum of two mark rows, viz., the mark rows of $\mathbf{D_4}(/\mathbf{C_2})$ + $\mathbf{D_4}(/\mathbf{C_{2v}})$, whence we write 1 under $\mathbf{C_2}$ and 1 under $\mathbf{C_{2v}}$ in the third row of the ICM and so on. The full ICM, *the heavens of Fujita*, c.f. FIG 35, is given below :

Table 14

Isomer-Count Matrix, ICM , which corresponds to the FPM (Table 12) and mark table. Table 3.

	C_1	C_4	C_2	C_s	C_4	C_{2v}	D_2	D_4
1	0	0	0	0	0	0	0	1^a
x	0	0	0	1^b	0	0	0	0
x^2	0	0	1^c	0	0	1^d	0	0
x^3	0	0	0	1^e	0	0	0	0
x^4	0	0	0	0	0	0	0	1^f

The letters a – f correspond to the structures shown in FIG 37.

* *

At this point a reader may not appreciate this relatively huge effort to solve (an almost) trivial problem (Something like killing a fly with an atomic bomb!). However as we introduce more examples, one will be convinced of the importance of this theory in all forms of combinatorial organic chemistry.

Example 34 Combinatoiral generation / symmetry classification of heterocyles

Find all derivatives which possess the general formula shown below. Compute isomer-count matrix and whence define (sub)-symmetries to these derivatives.

(six-membered ring of X atoms) ; X = { C, N } ... (153)

This is a D_6 point-group problem (six rotations and six reflections). The 12 operations are shown in FIG 29 and the table of multiplication is given below.

Table 15

Multiplication table of the D_6 group. Symbols are explaiend in FIG 29.

1st operation

2ed operation

	I	C_6	C_3	C_2	C_3'	C_6'	σ_1	σ_2	σ_3	σ_4	σ_5	σ_6
I	I	C_6	C_3	C_2	C_3'	C_6'	σ_1	σ_2	σ_3	σ_4	σ_5	σ_6
C_6	C_6	C_3	C_2	C_3'	C_6'	I	σ_4	σ_5	σ_6	σ_2	σ_3	σ_1
C_3	C_3	C_2	C_3'	C_6'	I	C_6	σ_2	σ_3	σ_1	σ_5	σ_6	σ_4
C_2	C_2	C_3'	C_6'	I	C_6	C_3	σ_5	σ_6	σ_4	σ_3	σ_1	σ_2
C_3'	C_3'	C_6'	I	C_6	C_3	C_2	σ_3	σ_1	σ_2	σ_6	σ_4	σ_5
C_6'	C_6'	I	C_6	C_3	C_2	C_3'	σ_6	σ_4	σ_5	σ_1	σ_2	σ_3
σ_1	σ_1	σ_6	σ_3	σ_5	σ_2	σ_4	I	C_3'	C_3	C_6'	C_2	C_6
σ_2	σ_2	σ_4	σ_1	σ_6	σ_3	σ_5	C_3	I	C_3'	C_6	C_6'	C_2
σ_3	σ_3	σ_5	σ_2	σ_4	σ_1	σ_6	C_3'	C_3	I	C_2	C_6	C_6'
σ_4	σ_4	σ_1	σ_6	σ_3	σ_5	σ_2	C_6	C_6'	C_2	I	C_3'	C_3
σ_5	σ_5	σ_2	σ_4	σ_1	σ_6	σ_3	C_2	C_6	C_6'	C_3	I	C_3'
σ_6	σ_6	σ_3	σ_5	σ_2	σ_4	σ_1	C_6'	C_2	C_6	C_3'	C_3	I

The multiplication properties lead to the following subgroups :

1. $\mathbf{C_1} = \{I\}$

2. $\mathbf{C_s} = \{I, \sigma_1\}$

3 $\mathbf{C_s'} = \{I, \sigma_4\}$

4. $\mathbf{C_2} = \{I, C_2\}$

5. $\mathbf{C_3} = \{I, C_3, C_3'\}$

6. $\mathbf{C_{2v}} = \{I, C_2, \sigma_1, \sigma_5\}$

7. $\mathbf{C_{3v}} = \{I, C_3, C_3', \sigma_4, \sigma_5, \sigma_6\}$

8. $\mathbf{C_6} = \{I, C_2, C_3, C_3', C_6, C_6'\}$

9. $\mathbf{C_3} = \{ I, C_3, C'_3, \sigma_1, \sigma_2, \sigma_3 \}$

10. $\mathbf{D_6}$

Other subgroups (such as $\mathbf{C}^*_s = \{ I, \sigma_2 \}$, $\mathbf{C}'_{2v} = \{ I, C_2, \sigma_3, \sigma_4 \}$ etc.) are conjugate subgroups and thus will not be included in the sequence of subgroups. Mark tables of $\mathbf{C_2}$, $\mathbf{C_S}$, $\mathbf{C}'_s$ and $\mathbf{C_{2v}}$ are already given (c.f. Tables 5, 7, 9 with a slightly different notations). We now give mark tables for other subgroups which can easily be derived:

Table 16

Mark table of $\mathbf{C_3}$

	$\mathbf{C_1}$	$\mathbf{C_3}$
C_3 (/C_1)	3	0
C_3 (/C_3)	1	1

Table 17

Mark Table of $\mathbf{C_6}$

	$\mathbf{C_1}$	$\mathbf{C_2}$	$\mathbf{C_3}$	$\mathbf{C_6}$
C_S (/C_1)	6	0	0	0
C_6 (/C_2)	3	3	0	0
C_6 (/C_3)	2	0	2	0
C_6 (/C_6)	1	1	1	1

Table 18

Mark Table of $\mathbf{D_3}$

	$\mathbf{C_1}$	$\mathbf{C_s}$	$\mathbf{C_3}$	$\mathbf{D_3}$
$\mathbf{D_3}$ (/$\mathbf{C_1}$)	6	0	0	0
$\mathbf{D_3}$ (/$\mathbf{C_s}$)	3	1	0	0
$\mathbf{D_3}$ (/$\mathbf{C_3}$)	2	0	2	0
$\mathbf{D_3}$ (/$\mathbf{D_3}$)	1	1	1	1

Table 17

Mark Table of $\mathbf{C_{3v}}$

	$\mathbf{C_1}$	$\mathbf{C'_s}$	$\mathbf{C_3}$	$\mathbf{C_{3v}}$
$\mathbf{C_{3v}}$ (/$\mathbf{C_1}$)	6	0	0	0
$\mathbf{C_{3v}}$ (/$\mathbf{C_s}$)	3	1	0	0
$\mathbf{C_{3v}}$ (/$\mathbf{C_3}$)	2	0	2	0
$\mathbf{C_{3v}}$ (/$\mathbf{C_{3v}}$)	1	1	1	1

Table 20

Mark Table of $\mathbf{D_6}$

	$\mathbf{C_1}$	$\mathbf{C_s}$	$\mathbf{C'_s}$	$\mathbf{C_2}$	$\mathbf{C_3}$	$\mathbf{C_{2v}}$	$\mathbf{C_{3v}}$	$\mathbf{C_6}$	$\mathbf{D_3}$	$\mathbf{D_6}$
$\mathbf{D_6}$(/$\mathbf{C_1}$)	12	0	0	0	0	0	0	0	0	0
$\mathbf{D_4}$(/$\mathbf{C_s}$)	6	2	0	0	0	0	0	0	0	0
$\mathbf{D_6}$ (/$\mathbf{C'_s}$)	6	0	2	0	0	0	0	0	0	0
$\mathbf{D_6}$(/$\mathbf{C_2}$)	6	0	0	6	0	0	0	0	0	0
$\mathbf{D_6}$(/$\mathbf{C_3}$)	4	0	0	0	4	0	0	0	0	0
$\mathbf{D_6}$(/$\mathbf{C_{2v}}$)	3	1	1	3	0	1	0	0	0	0
$\mathbf{D_6}$(/$\mathbf{C_{3v}}$)	2	0	2	0	2	0	2	0	0	0
$\mathbf{D_6}$(/$\mathbf{C_6}$)	2	0	0	2	2	0	0	2	0	0
$\mathbf{D_6}$(/$\mathbf{D_3}$)	2	2	0	0	2	0	0	0	2	0
$\mathbf{D_6}$(/$\mathbf{D_6}$)	1	1	1	1	1	1	1	1	1	1

Orbit – classification :

We label the six vertices, in sequence, 1,2,...,6 and operate by all 12 symmetry operations of $\mathbf{D_6}$, then count the number of fixed vertices under the effect of all subgroups (as we did for the square problem, eqn. (127). We obtain the following row of marks:

$$\begin{matrix} \mathbf{C_1} & \mathbf{C_s} & \mathbf{C'_s} & \mathbf{C_2} & \mathbf{C_3} & \mathbf{C_{2v}} & \mathbf{C_{3v}} & \mathbf{C_6} & \mathbf{D_3} & \mathbf{D_6} \\ (\ 6 & 2 & 0 & 0 & 0 & 0 & 0 & 0 & 0 & 0\ ) \end{matrix} \quad \ldots (154)$$

which corresponds to $\mathbf{D_6}(/C_s)$, Now we need the permutation representation, PR, of this coset which is necessary to obtain all subductions required to generate the USCI's . The PR's of this coset are tabulated below :

Table 21

Permutation representations of $\mathbf{D_6}$ $(/C_5)$

$\mathbf{D_6}(/C_s) = C_s + C_sC_2 + C_sC_3 + C_s C_3' + C_sC_6 + C_s C_6'$ PR

	1	2	3	4	5	6	
I	1	2	3	4	5	6	(1)(2)(3)(4)(5)(6)
C_2	2	1	6	5	4	3	(12)(36)(45)
C_3	3	6	4	1	2	5	(134)(265)
C_3'	4	5	1	3	6	2	(143)(256)
C_6	5	4	2	6	3	1	(153246)
C_6'	6	3	5	2	1	4	(164235)
σ_1	1	2	4	3	6	5	(1)(2)(34)(56)
σ_2	3	6	1	4	5	2	(4)(5)(13)(26)
σ_3	4	5	3	1	2	6	(3)(6)(14)(25)
σ_4	5	4	6	2	1	3	(15)(24)(36)
σ_5	2	1	5	6	3	4	(12)(35)(46)
σ_6	6	3	2	5	4	1	(16)(23)(45)

The next step is to subduce $\mathbf{D_6}$ $(/C_s)$ which is the orbit controlling substitution by all subgroups of $\mathbf{D_6}$. This is done below; c.f. eqns. (135) – (142) for the square problem.

1) $\mathbf{D_6}\ (/C_s) \downarrow C_1 \rightarrow$ (1)(2)(3)(4)(5)(6) $\rightarrow s_1^6$ ... (155)

2) $\mathbf{D_6}\ (/C_s \downarrow C_5 \rightarrow$ (1)(2)(3)(4)(5)(6) / (1)(2)(34)(56) (6 2)

$\rightarrow$ $2\ C_s\ (/C_1) + 2\ C_s\ (/C_5)$ $\rightarrow s_1^2\ s_2^2$ ... (156)

3) $\mathbf{D}_6\ (/\mathbf{C}_S) \downarrow \rightarrow \mathbf{C}'_s \rightarrow$

(1)(2)(3)(4)(5)(6) (15)(24)(36)

$\rightarrow$ (6 0)

$\rightarrow\ 2\ \mathbf{C}'_s\ (/\mathbf{C}_1) \rightarrow S_2^3$ ... (157)

4) $\mathbf{D}_6\ (/\mathbf{C}_S) \downarrow \rightarrow \mathbf{C}_2 \rightarrow$

(1)(2)(3)(4)(5)(6) (12)(36)(45)

$\rightarrow$ (6 0)

$\rightarrow\ 2\ \mathbf{C}_2\ (/\mathbf{C}_1) \rightarrow S_2^3$ ... (158)

5) $\mathbf{D}_6\ (/\mathbf{C}_S) \downarrow \rightarrow \mathbf{C}_3 \rightarrow$

(1)(2)(3)(4)(5)(6) (134)(265) (143)(256)

$\rightarrow$ (6 0)

$\rightarrow\ 2\ \mathbf{C}_3\ (/\mathbf{C}_1) \rightarrow S_2^3$ ... (159)

6) $\mathbf{D}_6\ (/\mathbf{C}_S) \downarrow\ \mathbf{C}_{2v} \rightarrow$

(1)(2)(3)(4)(5)(6) (12)(36)(45) (1)(2)(34)(56) (12)(35)(46)

$\rightarrow$ (6 0 2 0 0)

$\rightarrow\ \mathbf{C}_{2v}\ (/\mathbf{C}_1) + \mathbf{C}_{2v}\ (/\mathbf{C}_s) \rightarrow S_4\, S_2$ ... (160)

7) $\mathbf{D}_6\ (/\mathbf{C}_S) \downarrow\ \mathbf{C}_{3v} \rightarrow$

(1)(2)(3)(4)(5)(6) (134)(265) (143)(256) (15)(24)(36) (12)(35)(46) (16)(23)(45)

$\rightarrow$ (6 0 0 0)

$\rightarrow\ \mathbf{C}_{3v}\ (/\mathbf{C}_1) \rightarrow S_6$ ... (161)

8) $\mathbf{D_6}$ (/$\mathbf{C_S}$) ↓ $\mathbf{C_{X6}}$ →

(1)(2)(3)(4)(5)(6)
(12)(36)(45)
(134)(265)
(143)(256)
(153246)
(164235)

→ (6 0 0 0)

→ $\mathbf{C_6}$ (/$\mathbf{C_1}$) → S_6 ... (162)

9) $\mathbf{D_6}$ (/$\mathbf{C_S}$) ↓ $\mathbf{D_3}$ →

(1)(2)(3)(4)(5)(6)
(134)(265)
(143)(256)
(1)(2)(34)(56)
(4)(5)(13)(26)
(3)(6)(14)(25)

→ (6 2 0 0)

→ 2 $\mathbf{D_3}$ (/$\mathbf{C_S}$) → S_3^2 ... (163)

10) $\mathbf{D_6}$ (/$\mathbf{C_S}$) ↓ $\mathbf{D_6}$ →

All permutations In Table 21

→

(6 2 0 0 0 0 0 0 0 0) → $\mathbf{D_6}$ (/$\mathbf{C_S}$) → S_6 ... (164)

Now, we apply theorem of Pólya to the above USCI's with the weights :

$$w(C) = 1 \quad ; \quad w(N) = x$$

The coefficients of the resulting polynomials are collected in the following FP matrix:

Table 22

Fixed-point matrix, FPM , of the regular hexagon problem, ($\mathbf{D_6}$ point-group).

	C_1	C_s	C'_s	C_2	C_3	C_{2v}	C_{3v}	C_6	D_3	D_6
1	1	1	1	1	1	1	1	1	1	1
x	6	2	0	0	0	0	0	0	0	0
x^2	15	3	3	3	0	1	0	0	0	0
x^3	20	4	0	0	2	0	0	0	2	0
x^4	15	3	3	3	0	1	0	0	0	0
x^5	6	2	0	0	0	0	0	0	0	0
x^6	1	1	1	1	1	1	1	1	1	1

When the mark table of this point-group,Table 20, is applied to the above matrix, the IC matrix results :

Table 23

Isomer-count matrix, ICM, of the regular hexagon problem.

	C_1	C_s	C'_s	C_2	C_3	C_{2v}	C_{3v}	C_6	D_3	D_6
1	0	0	0	0	0	0	0	0	0	1^{ℓ}
x	0	1^b	0	0	0	0	0	0	0	0
x^2	0	1^c	1^g	0	0	1^i	0	0	0	0
x^3	1^a	1^d	0	0	0	0	0	0	1^k	0
x^4	0	1^e	1^h	0	0	1^j	0	0	0	0
x^5	0	1^f	0	0	0	0	0	0	0	0
x^6	0	0	0	0	0	0	0	0	0	1^m

The corresponding structures a, b, …, m are shown in FIG 38

Figure 38 (see page 188)

Example 35 The Tartaric Acids

Compute the fixed-point matrix and isomer-count matrix of the following substituted ethane drawn in its staggered conformation.

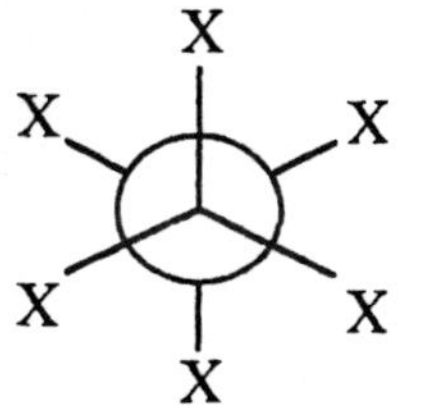

$$\text{(staggered ethane, six substituents X)} \quad ; \; X = \{ H, OH, COOH \} \qquad \ldots (165)$$

This is a $\mathbf{D_{3d}}$ point-group problem. Symmetry operations are modeled in FIG 2. The multiplication table is given below.

Table 24

Multiplication table of $\mathbf{D_{3d}}$

1st operation

		I	σ_d	σ_d'	σ_d''	C_3	C_3'	C_2	S_6	S_6'	σ_1	σ_2	σ_3
	I	I	σ_d	σ_d'	σ_d''	C_3	C_3'	C_2	S_6	S_6'	σ_1	σ_2	σ_3
	σ_d	σ_d	I	C_3	C_3'	σ_d'	σ_d''	σ_1	σ_3	σ_2	C_2	S_6'	S_6
	σ_d'	σ_d'	C_3'	I	C_3	σ_d''	σ_d	σ_2	σ_1	σ_3	S_6	C_2	S_6'
	σ_d''	σ_d''	C_3	C_3'	I	σ_d	σ_d'	σ_3	σ_2	σ_1	S_6'	S_6	C_2
2ed opn.	C_3	C_3	σ_d''	σ_d	σ_d'	C_3'	I	S_6'	C_2	S_6	σ_3	σ_1	σ_2
	C_3'	C_3'	σ_d'	σ_d''	σ_d	I	C_3	S_6	S_6'	C_2	σ_2	σ_3	σ_1
	C_2	C_2	σ_1	σ_2	σ_3	S_6'	S_6	I	C_3'	C_3	σ_d	σ_d'	σ_d''
	S_6	S_6	σ_2	σ_3	σ_1	C_2	S_6'	C_3'	C_3	I	σ_d'	σ_d''	σ_d
	S_6'	S_6'	σ_3	σ_1	σ_2	S_6	C_2	C_3	I	C_3'	σ_d''	σ_d	σ_d'
	σ_1	σ_1	C_2	S_6'	S_6	σ_2	σ_3	σ_d	σ_d''	σ_d'	I	C_3	C_3'
	σ_2	σ_2	S_6	C_2	S_6'	σ_3	σ_1	σ_d'	σ_d	σ_d''	C_3'	I	C_3
	σ_3	σ_3	S_6'	S_6	C_2	σ_1	σ_2	σ_d''	σ_d'	σ_d	C_3	C_3'	I

The sequence of representative subgroups (not including conjugata groups) is defined in FIG 2.

The orbit which governs substitution is a $\mathbf{D}_{3d}(/\mathbf{C}'_s)$ which when subduced by all subgroups leads to the following USCI's.

1. $\downarrow \mathbf{C}_1 \rightarrow S_1^6$

2. $\downarrow \mathbf{C}_s \rightarrow S_2^3$

3. $\downarrow \mathbf{C'}_s \rightarrow S_1^2 S_2^2$

4. $\downarrow \mathbf{C}_2 \rightarrow S_2^3$

5. $\downarrow \mathbf{C}_3 \rightarrow S_3^2$

6. $\downarrow \mathbf{C}_{2h} \rightarrow S_2 S_4$

7. $\downarrow \mathbf{D}_3 \rightarrow S_3^2$

8. $\downarrow \mathbf{C}_{3v} \rightarrow S_6$

9. $\downarrow \mathbf{S}_6 \rightarrow S_6$

10. $\downarrow \mathbf{D}_{3d} \rightarrow S_6$

We adopt the following weights:

$$w(H) = x \; ; \quad w(OH) = y \quad ; \quad w(COOH) = z \qquad \ldots (166)$$

When these weights are used to perform a Pólya-type substitution into the above USCI's we obtain the following fixed-point matrix :

Table 25

Fixed-point matrix of the staggered ethane problem.

	$\mathbf{C_1}$	$\mathbf{C_S}$	$\mathbf{C'_s}$	$\mathbf{C_2}$	$\mathbf{C_3}$	$\mathbf{C_{2h}}$	$\mathbf{D_3}$	$\mathbf{C_{3v}}$	$\mathbf{S_6}$	$\mathbf{D_{3d}}$
x^6	1	1	1	1	1	1	1	1	1	1
y^6	1	1	1	1	1	1	1	1	1	1
z^6	1	1	1	1	1	1	1	1	1	1
xy^5	6	0	2	0	0	0	0	0	0	0
x^5y	6	0	2	0	0	0	0	0	0	0
xz^5	6	0	2	0	0	0	0	0	0	0
x^5z	6	0	2	0	0	0	0	0	0	0
yz^5	6	0	2	0	0	0	0	0	0	0
y^5z	6	0	2	0	0	0	0	0	0	0
x^4yz	30	0	2	0	0	0	0	0	0	0
xy^4z	30	0	2	0	0	0	0	0	0	0
xyz^4	30	0	2	0	0	0	0	0	0	0
$x^2y^2z^2$	90	6	6	6	0	0	0	0	0	0
x^3y^2z	60	0	4	0	0	0	0	0	0	0
x^3yz^2	60	0	4	0	0	0	0	0	0	0
x^2y^3z	60	0	4	0	0	0	0	0	0	0
xy^3z^2	60	0	4	0	0	0	0	0	0	0
x^2yz^3	60	0	4	0	0	0	0	0	0	0
xy^2z^3	60	0	4	0	0	0	0	0	0	0
x^3y^3	20	0	4	0	2	0	2	0	0	0
x^3z^3	20	0	4	0	2	0	2	0	0	0
y^3z^3	20	0	4	0	2	0	2	0	0	0
x^4y^2	15	3	3	3	0	1	0	0	0	0
x^2y^4	15	3	3	3	0	1	0	0	0	0
x^4z^2	15	3	3	3	0	1	0	0	0	0
x^2z^4	15	3	3	3	0	1	0	0	0	0
y^4z^2	15	3	3	3	0	1	0	0	0	0
y^2z^4	15	3	3	3	0	1	0	0	0	0

There is a one-to-one correspondence between the subgroups of $\mathbf{D_{3d}}$ and those of the $\mathbf{D_6}$ (the preceeding example). C.f. Tables 15-19. Application of the mark table of $\mathbf{D_{3d}}$ (which is isomorphic with Table 19) generates the following ICM.

Table 26

The isomer-count matrix of the staggered ethane problem.

	C_1	C_s	C'_s	C_2	C_3	C_{2h}	D_3	C_{3v}	S_6	D_{3d}
x^6										1
y^6										1
z^6										1
xy^5			1							
x^5y			1							
xz^5			1							
x^5z			1							
yz^5			1							
y^5z			1							
x^4yz	2		1							
xy^4z	2		1							
xyz^4	2		1							
$x^2y^2z^2$	4	3	3	1						
x^3y^2z	4		2							
x^3yz^2	4		2							
x^2y^3z	4		2							
xy^3z^2	4		2							
x^2yz^3	4		2							
xy^2z^3	4		2							
x^3y^3	1		1				1			
x^3z^3	1		1				1			
y^3z^3	1		1				1			
x^4y^2		1	1			1				
x^2y^4		1	1			1				
x^4z^2		1	1			1				
x^2z^4		1	1			1				
y^4z^2		1	1			1				
y^2z^4		1	1			1				

Vacant positions in the above table are zeros.

We now verify some of the rows of the ICM. As an illustration the row which corresponds to $x^2 y^2 z^2$ corresponds to 4 isomers which possess $\mathbf{C_1}$ subsymmetry, 3 isomers which possess $\mathbf{C_s}$ subsymmetry, 3 isomers belonging to $\mathbf{C'_s}$ subgroup and one of $\mathbf{C_2}$ symmetry. These compounds are drawn in FIG 39

Figure 39 (see page 189)

The mark table of the $\mathbf{D_{3d}}$ point group is given below [isomorphic with Table 19].

Table 27

Mark table of $\mathbf{D_{3d}}$ point-group.

	C_1	C_s	C'_s	C_2	C_3	C_{2h}	D_3	C_{3v}	S_6	D_{3d}
$D_{3d}(/C_1)$	12	0	0	0	0	0	0	0	0	0
$D_{3d}(/C_s)$	6	2	0	0	0	0	0	0	0	0
$D_{3d}(/C'_s)$	6	0	2	0	0	0	0	0	0	0
$D_{3d}(/C_2)$	6	0	0	6	0	0	0	0	0	0
$D_{3d}(/C_3)$	4	0	0	0	4	0	0	0	0	0
$D_{3d}(/C_{2h})$	3	1	1	3	0	1	0	0	0	0
$D_{3d}(/D_3)$	2	0	2	0	2	0	2	0	0	0
$D_{3d}(/C_{3v})$	2	0	0	2	2	0	0	2	0	0
$D_{3d}(/S_6)$	2	2	0	0	2	0	0	0	2	0
$D_{3d}(/D_{3d})$	1	1	1	1	1	1	1	1	1	1

Now we sum the following rows multiplied by their appropriate factors:

4	$D_{3d}(/C_1)$	48	0	0	0	0	0	0	0	0	0
3	$D_{3d}(/C_s)$	18	6	0	0	0	0	0	0	0	0
3	$D_{3d}(/C'_s)$	18	0	6	0	0	0	0	0	0	0
1	$D_{3d}(/C_2)$	6	0	0	6	0	0	0	0	0	0
		(90	6	6	6	0	0	0	0	0	0)

... (167)

The vector given in (167) is the same as the vector which corresponds to $x^2y^2z^2$ in the FPM, Table 25. Naturally we could have obtained the multiplicities of the coset representations using the inverse of the mark table method, eqn. 112, however, we believe that finding these multiplicities *by inspection* may be both rewarding and exciting! Indeed after working out several examples it becomes an amusing experience. It tells us much about the real meaning of a mark: We can envisage a mark table as a **map of units** in the sense that each unit is a vector (the dimension of which = |SSG|) . For example the unit of the $\mathbf{C}_1$ isomer for the present problem is (2 1 0 0 0 0 0 0 0 0 0) and so on. The fixed-point vector given by eqn. (167) turns out to be resolvable into 4 $\mathbf{C}_1$ isomers, 3 $\mathbf{C}_s$ isomers, 3$\mathbf{C}'_s$ isomers and one $\mathbf{C}_2$ isomer. This type of achievement cannot be done by (simple !) theorem of Pólya. Analogously we can resolve the fixed-point vector corresponding to x^4yz into two isomers of $\mathbf{C}_1$ symmetry and one isomer of $\mathbf{C}'_s$ symmetry. These compounds are shown in FIG 40.

Figure 40 (see page 190)

In FIG 41 we show the 4 $\mathbf{C}_1$ isomers and the 2 C_S' isomers which corresponds to x^3y^2z.

Figure 41 (see page 191)

In FIG 42 we do the same for isomers of the type x^3y^3.

Figure 42 (see page 192)

Finally in FIG 43 we present the three x^4y^2 isomers classified according to their subgroups.

Figure 43 (see page 193)

FIG 39 may be particularly interesting to the organic chemist: the three (famous) stereoisomers of tartaric acid, viz. RS (≡ SR, *meso*-tartaric acid), RR and SS are naturally generated and their sub-symmetries assigned in this example. Here one must observe that this sub-symmetry classification is based on Newman projections or saw-hourse forms but *not* on the corresponding Fischer projections.

Example 36 Generation/Symmetry Classiciations of Heterocycles

Find all heterocycles which possess the following general formula :

$$\begin{array}{ccccc} X & - & X & - & X \\ | & & | & & | \\ X & - & X & - & X \end{array} \quad ; \; X = \{C, N\} \qquad \dots (167)$$

Assing appropriate symmetries to the resulting derivatives by computing their isomer-counting matrix.

This is a C_{2v} problem where we distinguish two orbits, viz.,

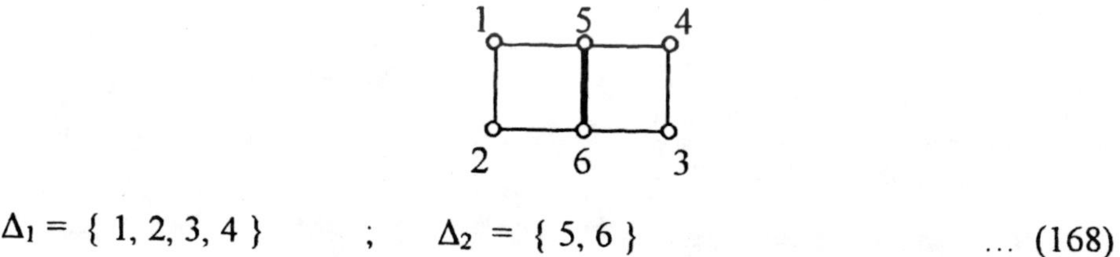

$$\Delta_1 = \{ 1, 2, 3, 4 \} \quad ; \quad \Delta_2 = \{ 5, 6 \} \qquad \dots (168)$$

FIG 2 models the C_{2v} point-group, its symmetry operations and sequence of subgrousp. The mark table is given by Table 9 where the subgroup C_4^2 may also be indicated as C_2 .

We adopt the following weights:

$$\left.\begin{array}{lll} w(C) = 1 & \text{for } \Delta_1 \text{ and } \Delta_2 \\ w(N) = x & \text{for } \Delta_1 \\ w(N) = y & \text{for } \Delta_2 \end{array}\right\} \qquad \dots (169)$$

It is easy to show that Δ_1 is controled by C_{2v} ($/C_1$) while Δ_2 is governed by C_{2v} ($/C_s$)

- Subductions of C_{2v} ($/C_1$) generate the following USCI's :

$$\downarrow \mathbf{C}_1 \rightarrow S_1^4 \rightarrow (1+x)^4 \quad \ldots (170)$$

$$\downarrow \mathbf{C}_2 \rightarrow S_2^2 \rightarrow (1+x^2)^2 \quad \ldots (171)$$

$$\downarrow \mathbf{C}_s \rightarrow S_2^2 \rightarrow (1+x^2)^2 \quad \ldots (172)$$

$$\downarrow \mathbf{C}_s' \rightarrow S_2^2 \rightarrow (1+x^2)^2 \quad \ldots (173)$$

$$\downarrow \mathbf{C}_{2v} \rightarrow S_4 \rightarrow (1+x^4) \quad \ldots (174)$$

Where the generating functions are obtained by the usual application of Pólya's theorem.

- Subduction of $\mathbf{C}_{2v}$ (/$\mathbf{C}_s$) generate the following USCI's and the corresponding polynomials:

$$\downarrow \mathbf{C}_1 \rightarrow S_1^2 \rightarrow (1+y)^2 \quad \ldots (175)$$

$$\downarrow \mathbf{C}_2 \rightarrow S_2 \rightarrow (1+y^2) \quad \ldots (176)$$

$$\downarrow \mathbf{C}_s \rightarrow S_1^2 \rightarrow (1+y)^2 \quad \ldots (177)$$

$$\downarrow \mathbf{C}_s' \rightarrow S_2 \rightarrow (1+y^2) \quad \ldots (178)$$

$$\downarrow \mathbf{C}_{2v} \rightarrow S_2 \rightarrow (1+y^2) \quad \ldots (179)$$

Now we multiply the polynomials obtained from Δ_1, eqns. (170)-(174), times those obtained from Δ_2, eqns. (175)-(179) and the coefficients are used to construct the following fixed-point matrix.

Table 28

Fixed-point matrix of the bicyclobutadiene molecular graph problem.

	C_1	C_2	C_s	C_s'	C_{2v}
1	1	1	1	1	1
x	4	0	0	0	0
y	2	0	2	0	0
y^2	1	1	1	1	1
x^2	6	2	2	2	0
xy	8	0	0	0	0
x^3	4	0	0	0	0
x^2y	12	0	4	0	0
xy^2	4	0	0	0	0
x^4	1	1	1	1	1
x^3y	8	0	0	0	0
x^2y^2	6	2	2	2	0
x^4y	2	0	2	0	0
x^3y^2	4	0	0	0	0
x^4y^2	1	1	1	1	1

And by combining the appropriate rows from the mark table, Table 9, with appropriate multiplicities we arrive at the desired isomer-counting matrix:

Table 29

Isomer-counting matrix of all heterocycles derived from the molecular graph of bicyclobutadiene.

	C_1	C_2	C_s	C_s'	C_{2v}
1	.	.	.	.	.
x	$*^{2}$	.	.	.	.
y	.	.	$*^{3}$	.	.
y^2	.	.	.	.	$*^{4}$
x^2	.	$*^{5}$	$*^{6}$	$*^{7}$	.
xy	$*^{8}\ *^{9}$	.	.	.	.
x^3	$*^{10}$	.	.	.	.
x^2y	$*^{11}\ *^{12}$	.	$*^{13}\ *^{14}$	.	.
xy^2	$*^{15}$	.	.	.	.
x^4	.	.	.	.	$*^{16}$
x^3y	$*^{17}\ *^{18}$	.	.	.	.
x^2y^2	.	$*^{19}$	$*^{20}$	$*^{21}$	.
x^4y	.	.	$*^{22}$	.	.
x^3y^2	$*^{23}$	.	.	.	.
x^4y^2	.	.	.	.	$*^{24}$

The 24 derivatives are labeled as 1,2,.., 24 , are shown in FIG 44 along with their sub-symmetry classification.

Figure 44 (see page 194)

Example 37

Find the isomer-count mjatrix of the following substituted trishomocubane graph.

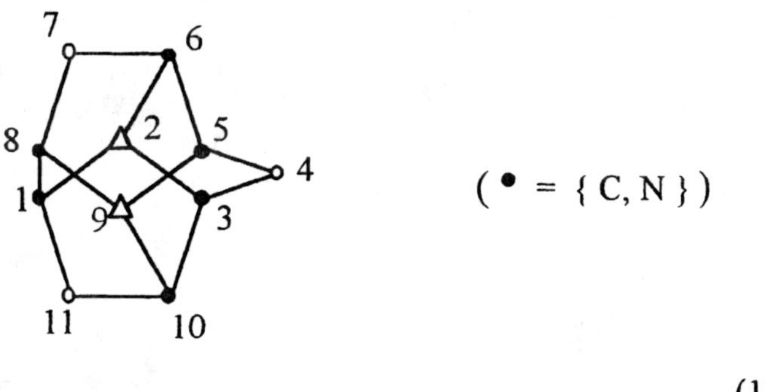

... (180)

Trishomocubane is a caged compound which belongs to the $\mathbf{D}_3$ point-group [c.f. FIG 1]. The compound can be envisaged from the fusion of six equivalent cyclopentane rings (or three norbornanes). The parent (unsubstituted graph of this molecule has three C_2 axes of symmetry each passing through a methylene carbon and the center of an opposite edge: One C_2 axis passes through vertex 4 and edge joining vertices 1 and 8, Another C_2 axes passes through vertex 7 and the bond joining vertices 3 and 10. Finally a third C_2 passing through vertex 11 and the bond between vertices 5 and 6. The molecular also has two C_3 axes, each passes through vertices 1 and 9. Inspection of the graph shown in eqn. (180) reveals three orbits, viz.,

$\Delta_1 = \{ 4, 7, 11\}$; (colored as open circles) ... (181)

A subset of bivalent vertices.

$\Delta_2 = \{ 1, 3, 5, 6, 8, 10\}$; (colored as solid circles) ... (182)

A subset of trivalent vertices each one is adjacent to two trivalent vertices and one bivalent vertex.

$\Delta_3 = \{2,9\}$; (colored as open triangles) ... (183)

A subset of trivalent vertices, each vertex of which is surrounded by three trivalent vertices.

The mark table of $\mathbf{D}_3$ is given by Table 30:

Table 30

Mark table of $\mathbf{D_3}$ point-group

	$\mathbf{C_1}$	$\mathbf{C_2}$	$\mathbf{C_3}$	$\mathbf{D_3}$
$\mathbf{D_3(/C_1)}$	6	0	0	0
$\mathbf{D_3(/C_2)}$	3	1	0	0
$\mathbf{D_7(/C_3)}$	2	0	2	0
$\mathbf{D_3(/D_3)}$	1	1	1	1

[C.f. Table 18, indeed $\mathbf{C_2} \equiv \mathbf{C_4}^2 \equiv \mathbf{C_s}$]

Now to define the coset representation which controls the orbit of substitution, Δ_2 in this case, eqn (182), we apply the elements of symmetry of $\mathbf{D_3}$ to the vertices of Δ_2 and count the number of fixed vertices under the effects of all subgroups: the rows of the mark table of the full group (Table 30). These operations are shown below:

			$\mathbf{C_1}$	$\mathbf{C_2}$	$\mathbf{C_3}$	$\mathbf{D_3}$
I	$\rightarrow$	(1)(6)(5)(3)(10)(8)	√	√	√	√
C_2	$\rightarrow$	(18)(6 10) (3 5)		√		√
C_2'	$\rightarrow$	(3 10) (15) (68)				√
C_2''	$\rightarrow$	(5 6) (1 10) (38)				√
C_3	$\rightarrow$	(6 1 3) (5 8 10)			√	√
C_3'	$\rightarrow$	(6 3 1) (5 10 8)			√	√

$\mathbf{C_1}$	$\mathbf{C_2}$	$\mathbf{C_3}$	$\mathbf{D_3}$
(6	0	0	0)

... (184)

Comparison of the vector generated in eqn. (184) with mark table, Table 29, indicates that the orbit which controls substitution in Δ_2 is $\mathbf{D_3}$ (/$\mathbf{C_1}$) type. This is the CR which must be subduced by the four subgroups of $\mathbf{D_3}$. The resulting USCI's are outlined below together with the corresponding generating functions adopting the weights:

$$w(C) = x \quad ; \qquad w(N) = y \qquad \text{... (185)}$$

$$\downarrow \mathbf{C_1} \rightarrow s_1^6 \rightarrow (x+y)^6 \qquad \text{... (186)}$$

$$\downarrow \mathbf{C_2} \rightarrow s_2^3 \rightarrow (x^2+y^2)^3 \qquad \text{... (187)}$$

$$\downarrow \mathbf{C_3} \rightarrow s_3^2 \rightarrow (x^3+y^3)^2 \qquad \text{... (188)}$$

$$\downarrow \mathbf{D_3} \rightarrow s_6 \rightarrow (x^6+y^6) \qquad \text{... (189)}$$

The above polynomials when expanded generate the following FP matrkx:

Table 31

Fixed-point matrix of the trishomocubane problem.

	C₁	**C₂**	**C₃**	**D₃**
x^6	1	1	1	1
y^6	1	1	1	1
xy^5	6	0	0	0
x^5y	6	0	0	0
x^2y^4	15	3	0	0
x^4y^2	15	3	0	0
x^3y^3	20	0	2	0

When the mark table is applied into Table 31 we obtain the desired isomer-count matrix. shown below:

Table 32

Isomer-count matrix of the trishomocubane graph.

	C_1	C_2	C_3	D_3
x^6				$*^{1}$
y^6				$*^{2}$
xy^5	$*^{3}$			
x^5y	$*^{4}$			
x^2y^4	$*^{5}$	$*^{6}$ $*^{7}$ $*^{8}$		
x^4y^2	$*^{9}$	$*^{10}$ $*^{11}$ $*^{12}$		
x^3y^3	$*^{13}$ $*^{14}$ $*^{15}$		$*^{16}$	

Where each star corresponds to a row from the mark table. The labelings of the stars correspond to the heterocycles derived from this caged molecule. These are drawn in FIG 45.

Figure 45 (see page 195)

3.2.7. A "re-visit" to nmr

We can use Fujita's theory of orbit classification to study nmr of a given configuration according to the following steps:

1) Define point-group of graph and find all its subgroups.
2) Compute mark table of that point-group.
3) Apply all symmetry operations to the vertices of the graph.
4) Calculate the number of fixed vertices under the operations of each subgroup and whence find the row of marks.
5) Extract from the mark table the row (or rows) which when summed together (after being multiplied by proper multiplication factors) generate the row of marks generated in step 4).
6) The number of cost representations extracted from the mark table in step 5) = the number of (magnetically) equivalent nuclei.

Example 38 ^{13}C nmr of naphthalene

This is a $\mathbf{C_{2v}}$ point-group problem, c.f. FIG 2 .

The labelings of the vertices of the molecular graph of naphthalene are shown in FIG 34. Now we find the number of vertices which remain invariant under the symmetry operations of the subgroups of $\mathbf{C_{2v}}$. This is computed below:

			$\mathbf{C_1}$	$\mathbf{C_2}$	$\mathbf{C_s}$	$\mathbf{C_s'}$	$\mathbf{C_{2v}}$
I	$\rightarrow$	(1)(2)(3)(4)(5)(6)(7)(8)(9)(10)	√	√	√	√	√
C_2	$\rightarrow$	(16)(27)(38)(49)(5 10)		√			√
σ_v	$\rightarrow$	(5)(10)(19)(28)(37)(46)			√		√
σ_h	$\rightarrow$	(14)(23)(10 5)(96)(87)				√	√
			(10	0	2	0	0)

... (190)

This mark row is consistent with $2\mathbf{C_{2v}}\,(/\mathbf{C_1}) + \mathbf{C_{2v}}\,(/\mathbf{C_s})$ as can be anticipated from the mark row of $\mathbf{C_{2v}}$, Table 10. This means that there are three orbits, i.e., three sets of magnetically equivalent nuclei such that two orbits are being controlled by $\mathbf{C_{2v}}(/\mathbf{C_1})$ and one orbit which is controlled by $\mathbf{C_{2v}}\,(/\mathbf{C_s})$. This conclusion is vertified as follows:

* $\Delta_1 = \{ 2, 3, 8, 7 \}$, C.f. FIG 34 for vertex-labelings.

			$\mathbf{C_1}$	$\mathbf{C_2}$	$\mathbf{C_s}$	$\mathbf{C_s'}$	$\mathbf{C_{2v}}$
I	$\rightarrow$	(2)(3)(8)(7)	√	√	√	√	√
C_2	$\rightarrow$	(27)(38)		√			√
σ_v	$\rightarrow$	(23)(87)			√		√
σ_h	$\rightarrow$	(28)(37)				√	√
		Mark row : (	4	0	0	0	0)

... (191)

The row of marks given by eqn. (191) corresponds to C_{2v} (/C_1), c.f. mark table, Table 10

			C_1	C_2	C_s	C_s'	C_{2v}
I	→	(1)(4)(6)(9)	√	√	√	√	√
C_2	→	(16)(49)		√			√
σ_v	→	(19)(46)			√		√
σ_h	→	(14)(96)				√	√
		Mark row :	(4	0	0	0	0)

... (192)

This mark row is also consistent with the CR C_{2v} (C_1/C_1).

* Δ_3 = { 5 10 }

			C_1	C_2	C_s'		C_{2v}
I	→	(5)(10)	√	√	√	√	√
C_2	→	(5 10)					√
σ_v	→	(5) (10)		√	√		√
σ_h	→	(5 10)				√	√
		Mark row :	(2	0	2	0	0)

... (193)

The above mark row corresponds to C_{2v} (/C_s).

Therefore the ^{13}C nmr of naphthalene involves three sets of magnetically equivalent nuclei: two sets are governed by C_{2v} (/C_1) and one being controlled by C_{2v} (/C_s). The formula of Balasubramanian, eqn. (98) computes three orbits while the theory of Fujita classifies these orbits according to their coset representations. In FIG 34 these orbits are colored differentially.

Bibliography

1. N. Trinajstic, Chemical graph theory,CRC press, Boca Raton, Fl. Vol. 1,2 (1992).
2. I. Gutman and S.J. Cyvin, Introduction to the theory of benzenoid hydrocarbons Springer-Verlag, Berlin (1989).
3. B.S. Tsukerblat,Group theory in chemistry and spectroscopy, Academic press, London (1994).
4. S. Fujita, Symmetry and combinatorial enumeration in chemistry,Springer-Verlag,Berlin (1991).
5. W. Burnside, Theory of groups of finite order, 2nd ed., Cambridge Univ. Press,Cambridge (1911).
6. This observation has been stated by professor C.A.Mead on his comment on Fujita's work; C.A. Mead, J. Am. Chem.Soc., **114**(10), 4018-4019 (1992).
7. Conjugate subgroups are defined as follows: let $\mathbf{H} = \{ g_1, g_2, .., g_n \}$ and $\mathbf{H}' = g_1'$, $g_2' \ldots, g_n' \}$. Then **H** and $\mathbf{H}'$ are called conjugated if $\{ g_1, g_2, .., g_n \} = s\{ g_1', g_2', \ldots, g_n' \} S^{-1}$ where $s \in \mathbf{G}$, i.e., is an element of the parent group **G**. See, e.g., R.F. Flurry, Jr, Symmetry groups, Theory and chemical applications, Prentice-Hall, Inc., Englewood Cliffs, New Jersey, p. 33 (1980).
8. S. Fujita, Systematic classification of molecular symmetry by subduction of coset representations, Bull. Chem. Soc. Japan, **63**, 315-327 (1990).
9. S. Fujita, Subduction of coset representations. An application to enumeration of chemical structures, Theor. Chim. Acta, **76**, 247-268 (1989).
10. G. R. Underwood and B.Ramamoorthy , Chemical Studies of Caged Compounds , The Synthesis of pentacyclo[$6,3,0,0^{2,6},0^{3,10},0^{5,9}$] undecane , Tetrahedron lett. **47** , pp4125-4127(1970).
11. K. Balasubramanian , Application of Combinatorics and Graph theory to spectroscopy and quantum chemistry , Chem. Rev. , **85**(6) , pp 602-603(1985).

Chapter 4

4 Modeling Symmetry and Combinatorial Enumeration In Chemistry

4.1 Modeling Coset Representation of a Group

4.1.1The Identity Representation , G(/G)

The model in this particular case is a graph which remains invariant under the operation of any $g \in \mathbf{G}$, where g is an element of symmetry such that the number of vertices in its orbit, $\Delta = |\mathbf{G}|$. For example the **T** group has 12 symmetry elements, i.e., $|\mathbf{T}| = 12$ and whence the molecular graph of methane (4H atoms, $|\Delta| = 4$) is not a suitable model for the **T(/T)** representation. On the other hand adamantane molecular graph contains 12 H atoms along the periphery of the molecule which make it suitable for modeling **T(/T)** . **Fig**.1-3 shows models for the identity representations of 15 points groups (which are of chemical importance). One observes that the graphs shown in **fig**.1 remain invariant under symmetry operations of the relevant groups **only**. This is not exactly trivial : consider e.g., a model for $\mathbf{C_3(/C_3)}$ [where $\mathbf{C_3} = \{1, C_3, C_3'\}$] obviously an equilateral triangle remains invariant under any of the operations of $\mathbf{C_3}$, furthermore $|\Delta| = 3 = |\mathbf{C_3}|$. However such a model remains **also** fixed under 3 mirror planes (one passing through each of the three vertices of the triangle) and therefore cannot be adopted to model this representation . The correct model is shown in **Fig**.1.

4.1.2 Other Coset Representations

Firstly we observe that the other CR's amount to "reduction" (or indeed **subduction**) of symmetry of the identity graph. For example the molecular graph of cyclobutane (c.f. **Fig**.1) belongs to $\mathbf{D_4}$ group while the molecular graph of cyclobutane in which two nonadjacent carbons have been replaced by two N atoms belongs to $\mathbf{C_{2v}}$, a subgroup of $\mathbf{D_4}$. One way to reduce the symmetry of the parent graph (i.e., the model of the identity representation) is to perform the appropriate **substitution** on some of the vertices of the orbit of **G(/G)** model graph. The mathematical synonym of substitution is **coloring** . We adopt the following convention : all the vertices of the parent graph are open (≡ white) while for the other graphs (which model other CR's) some of the vertices will be closed (≡ black) while others open. But how can we perform the coloring operation so that the resulting graphs generate the same algebra of the CR it represents and whence produce its row of marks? In other words : suppose we wish to find a model graph for the CR : $\mathbf{G(/G_i)}$, then we start from the (

uncolored) graph of **G(/G)** (shown in **Fig**.1) and then we apply all the elements of symmetry of the subgroup $\mathbf{G}_i$. Now each $g \in \mathbf{G}_i$ generates a certain permutation of the vertices (of Δ) which can be expressed in cyclic notation. Typically one might have :

$$(\upsilon_a \quad \upsilon_b)(\upsilon_c \quad \upsilon_d \quad \upsilon_e \quad \upsilon_f)(\ldots) \qquad \ldots (194)$$

where υ_a is the ath vertex of Δ and so on . The above (and such equations) are then "translated" , so - to - speak into "color identities" . Eqn.(194), e.g.,leads to the following vertex - color identities :

$$\upsilon_a = \upsilon_b$$

$$\upsilon_c = \upsilon_d = \upsilon_e = \upsilon_f \qquad \ldots (195)$$

Eqns.(195) are understood to mean that the color of υ_a = the color of υ_b and so on .

Arbitrarily we select the number of black vertices to be a minimum.
Fig.46 shows the two nonaxial groups : $\mathbf{C}_1$ and $\mathbf{C}_s$.

Figure 46 (see page 196)

The molecule CHFClBr belongs to $\mathbf{C}_1$ ={I} , with trivially one CR :$\mathbf{C}_1(/\mathbf{C}_1)$, a single vertex . $\mathbf{C}_s$ has in addition to I a σ plane and a model for it would be the molecule of SO_2BrF where the σ is the plane of SFBr , i.e., the plane of the paper. The two CR's are shown in **Fig**.46 .
Fig.47. contains some of the $\mathbf{C}_n$ groups (cyclic groups) viz., $\mathbf{C}_2$,$\mathbf{C}_3$and $\mathbf{C}_4$.

The molecule of 1,1-dichloroethylene in which the CH_2 and CCl_2 planes are rotated through an angle $\alpha \neq \frac{1}{2}\pi$ is an example of a molecule which belongs to $\mathbf{C}_2$. (when α

= ½π , σ planes are formed and a C_{2v} point group is generated). Tetraphenylcyclobutadiene in which the planes of the phenyl groups are orthogonal to the plane of the paper (while the cyclobutadiene ring is in the plane of the paper) represents a $\mathbf{C_4}$ point group . Models of the CR's of these $\mathbf{C_n}$ groups are portrayed in **Fig**.47.

Figure 47 (see page 197)

Fig.48 presents examples of the **dihedral** point groups $\mathbf{D_2}$, $\mathbf{D_3}$ and $\mathbf{D_4}$.

Figure 48 (see page 198-199)

These $\mathbf{D_n}$ groups contain as symmetry elements the vertical C_n axis and n horizontal C_2 axes intersecting at angles π/n. Biphenyl belongs to $\mathbf{D_2}$ while the gauche conformation of ethane possesses a $\mathbf{D_3}$ structure.

Point groups $\mathbf{C_{2v}}$ and $\mathbf{C_{3v}}$ are modeled in **Fig**.49.

Figure 49 (see page 200)

Water is a prototype example of $\mathbf{C_{2v}}$ point group but it cannot model all of its (five) coset representations. The molecular graph of bicyclohexatriene whose orbit of substitution is the four peripheral vertices turns out to be the model of choice for this point group. Similarly , although PCl_3 belongs to $\mathbf{C_{3v}}$ its orbit does not have enough vertices for appropriate modeling of all possible CR's. A convenient model is the molecular graph in which three regular pentagons are fused together so that all share one (central) vertex and any two share one edge. Here the orbit will be the set of all bivalent vertices, i.e. $|\Delta| = 6 = |\mathbf{C_{3v}}|$. This model represents the π - system of acepentylene. **Fig**.50 models $\mathbf{C_{3h}}$ point group. This point group is formed by the addition of a horizontal plane to a C_n system. The molecule of boric acid exemplifies this group.

Figure 50 (see page 201)

In **Fig**.51 we consider the $\mathbf{D_{3h}}$ system formed by addition of a horizontal plane to the system of axes of the corresponding dihedral ($\mathbf{D_n}$) point-group.

Figure 51 (see page 202)

The $\mathbf{D_{3h}}$ system is modeled by the molecular graph of benzene in which three cyclobutane rings are condensed around the periphery of the hexagon .This way the horizontal plane is defined between the H atoms above and below the plane of the paper. There are 6 H atoms above and 6 H atoms below this plane making a total of 12 vertices which constitute the orbit of substitution = $|\mathbf{D_{3h}}|$. **Fig**.52 illustrates two examples of the $\mathbf{D_{nd}}$ point group ; viz., $\mathbf{D_{2d}}$ and $\mathbf{D_{3d}}$.

Figure 52 (see page 203-204)

This system is formed by attaching vertical diagonal planes to the axis system of $\mathbf{D_n}$. There are 8 elements of symmetry in $\mathbf{D_{2d}}$ and whence the 4 H atoms of allenes are not suitable for defining an orbit of substitution. A suitable model is shown in **Fig**.52 and similarly for $\mathbf{D_{3d}}$ .The $\mathbf{S_4}$ point group is modeled in **Fig**.53 and finally the cubic , **T** , group is modeled by the adamantane molecular graph in **Fig**.54.

Figure 53 (see page 205)

Figure 54 (see page 206)

4.1.3 An Illustration : The set of homomers

The $\mathbf{C_{3v}}$ point group : [see **Figs**. 2 and 49]

The $\mathbf{C_{3v}}$ point group has six elements of symmetry , viz.,

$$C_{3v} = \{I, C_3, C_3', \sigma_v, \sigma_v', \sigma_v''\} \quad \ldots (196)$$

These symmetry elements and their operations are illustrated in **Fig**. 2. The sequence of subgroups and their definitions are given below:

$$SSG = \mathbf{C}_1 ;\ \mathbf{C}_s = \{I, \sigma_v\} ;\ \mathbf{C}_3 = \{I, C_3, C_3'\} \qquad \text{... (197)}$$

Our model for $\mathbf{C}_{3v}(/\mathbf{C}_{3v})$ is shown by the following graph where the labeled vertices constitute the orbit of $\mathbf{C}_{3v}(/\mathbf{C}_{3v})$:

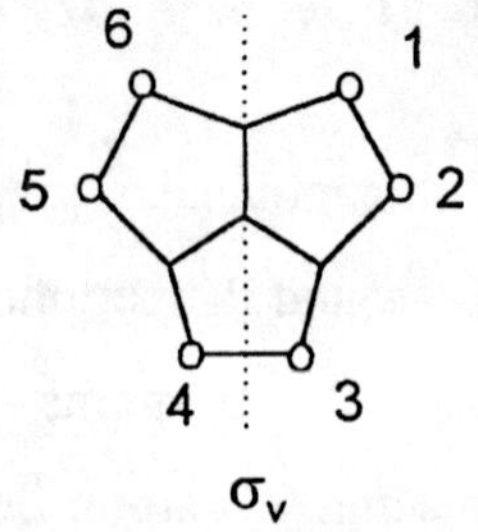

6 1
5 2
4 3
σ_v

...(198)

$$\text{I.e.,} = \Delta(\mathbf{C}_{3v}(/\mathbf{C}_{3v})) = \{\upsilon_1, \upsilon_2, \ldots, \upsilon_6\} \qquad \text{... (199)}$$

The above model stays invariant under all g's $\in \mathbf{C}_{3v}$ .Suppose now we wished to model the CR : $\mathbf{C}_{3v}(/\mathbf{C}_s)$: Then , we apply $\mathbf{C}_s = \{I, \sigma_v\}$ to the above graph . We observe that I is not significant because it leaves any thing fixed (by definition!) but σ_v generates the following permutation:

$$(16)(25)(34) \qquad \text{... (200)}$$

Which implies the following color equalities of the vertices:

$$1 = 6 \ ; 2 = 5 ;\ 3 = 4 \qquad \text{... (201)}$$

If , then , one selects 1 = 6 = 2 = 5 = white ; 3 = 4 = black (keeping the number of black vertices minimum) , the following colored configuration is obtained .

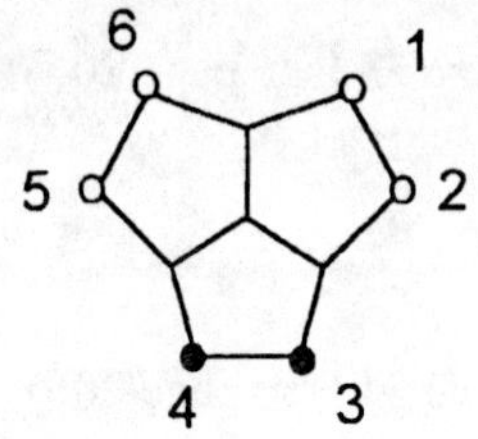

6 1
5 2
4 3
h_1

... (202)

Naturally the above colored graph remains invariant **only** under the elements of $\mathbf{C}_s$ and we emphasize here the word **only** !(The "uncolored" graph also remains fixed under $\mathbf{C}_s$ **as well as other groups** and whence it does not work as a model for $\mathbf{C}_{3v}(/\mathbf{C}_s)$ representation) . But now we must look for all other (colored) configurations which are equivalent to the configuration labeled h_1 (in eqn.202). At this point we recall the **chemical** term for **equivalent configurations** is **homomers** [1]. Indeed two (positional) isomers are called **homomeric** if :

i) They have orbits of the same type , and

ii) They are superimposable by proper rotations.

Thence the safest way to find all other homomers of h_1 is to apply all $g \in \mathbf{C_{3v}}$ to h_1 and draw the resulting (necessarily) equivalent configurations. This shown in **Fig**.55 where , e.g., h_1 is transformed into h_2 by **either** of the symmetry elements in the braces , {}'s ; i.e. either by C_3 **or** by σ_v'' and similarly for other transformations.

Figure 55 (see page 207)

The set $\{h_1, h_2, h_3\}$ will be given the symbol $\mathcal{H}$ $[\mathbf{C_{3v}}(/\mathbf{C_s})]$ or simply $\mathcal{H}$. Here, we observe that a given CR, which is **indeed** an abstract concept, has been replaced by a **set of graphs**. We now demonstrate that:

i) There is a **one-to-one** ; **onto** correspondence [2] (i.e., a **bijection**) between the elements of $\mathcal{H}$ and the coset decomposition of $\mathbf{C_{3v}}$ by $\mathbf{C_s}$,

ii) $\mathcal{H}$, indeed, generates the same marks as obtained algebraically. I.e., we demonstrate the equivalence between **algebraic** and **graphical** generation of marks :

Firstly we observe the following equalities :

$$\mathbf{C_{3v}/C_s} = \{\mathbf{C_s}, \mathbf{C_s}C_3', \mathbf{C_s}C_3\} = \{\{I,\sigma_v\}, \{C_3', \sigma_v''\}, \{C_3, \sigma_v'\}\} ; \quad \dots (203)$$

$$\mathcal{H}\,[\mathbf{C_{3v}}(/\mathbf{C_s})] = \{h_1, h_2, h_3\} \quad \dots (204)$$

Now, we wish to see whether there is a bijection between:

$h_1 \leftrightarrow \{I, \sigma_v\} \equiv \{\mathbf{C_s}\}$;

$h_2 \leftrightarrow \{C_3', \sigma_v''\} \equiv \{\mathbf{C_s}C_3'\}$;

$h_3 \leftrightarrow \{C_3, \sigma_v'\} \equiv \{\mathbf{C_s},C_3\}$ ... (205)

So we form tables of permutations generated by both sets (given by eqns.(203) and (204)) when operated upon by all g's $\in \mathbf{C_{3v}}$. The two sets indeed generate the same permutations as shown in tables 33 and 34. For example the second row of **Table 34** is concluded in **Fig**.56.

Figure 56 (see page 208)

Table 33

Permutations generated by the coset representation $C_{3v}(/C_s)$.(The coset - partitions of C_{3v} are labeled as 1,2 and 3)

$C_{3v}(/C_s) = C_s + C_sC_{3'} + C_sC_3$

	$\{I, \sigma_v\}$	$\{C_{3'}, \sigma_v''\}$	$\{C_3, \sigma_v'\}$	
	1	2	3	
Symmetry Element				Permutation
I	1	2	3	(1)(2)(3)
C_3'	2	3	1	(123)
C_3	3	1	2	(132)
σ_v	1	3	2	(1)(23)
σ_v'	3	2	1	(2)(13)
σ_v''	2	1	3	(3)(12)

Table 34

Permutations generated from the set of colored graphs (homomers) which represent $C_{3v}(/C_s)$, See eqn.(207) and **Fig**. 56.

(The homomers are labeled as 1,2 and 3)

$\mathcal{H}[C_{3v}(/C_s)]$	$\{h_1$	h_2	$h_3\}$	
Element of symmetry	1	2	3	Permutation
I	1	2	3	(1)(2)(3)
C_3'	2	3	1	(123)
C_3	3	1	2	(132)
σ_v	1	3	2	(1)(23)
σ_v'	3	2	1	(2)(13)
σ_v''	2	1	3	(3)(12)

Either tables 33 or 34 leads to the following row of marks:

	C_1	C_s	C_3	C_{3v}
$I \rightarrow$ (1) (2) (3)	✔	✔	✔	✔
$C_3' \rightarrow$ (123)			✔	✔
$C_3 \rightarrow$ (132)			✔	✔
$\sigma_v \rightarrow$ (1) (23)		✔		✔
$\sigma_v' \rightarrow$ (2) (13)				✔
$\sigma_v'' \rightarrow$ (3) (12)				✔

$$\begin{matrix} C_1 & C_s & C_3 & C_{3v} \\ (\;3 & 1 & 0 & 0\;) \end{matrix} \quad \ldots (206)$$

We may then say that $\mathcal{H}$ $[C_{3v}(/C_s)] \equiv \{h_1,h_2,h_3\}$ **models** the CR: $C_{3v}(/C_s)$: while $C_{3v}(/C_s)$ generates the marks **algebraically** using the multiplication properties of symmetry elements , the set of homomers generates the very cyclic structures **graphically** , by **only** considering rotation/reflection properties of simple graphs.

An alternative approach to obtain the marks is to draw all the homomers of a given **H**[G(/G$_i$)] and record below each homomer the subgroup(s) (selected from the SSG) under which it remains fixed . For the case at hand, $C_{3v}(/C_s)$, this is shown below:

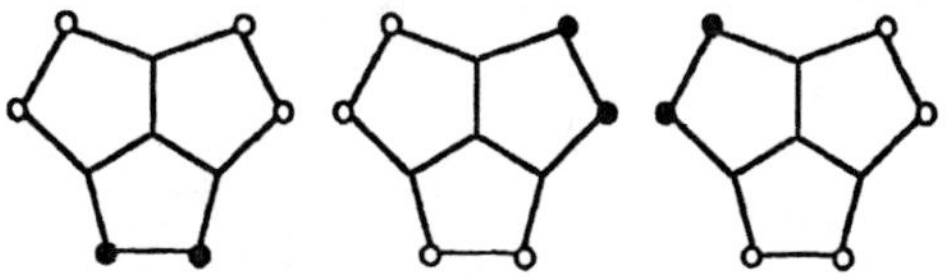

h$_1$ h$_2$ h$_3$

$$\begin{matrix} C_1 & C_1 & C_1 \rightarrow 3 \\ C_s & & \rightarrow 1 \end{matrix} \Big\}$$

...(207)

Which generates the same row of marks as given by eqn.(206). Indeed the row of marks which corresponds to $C_{3v}(/C_s)$ is best easily envisaged if modeled in the following form:

	C_1	C_s	C_3	C_{3v}
$C_{3v}(/C_s)$	{ h$_1$ h$_2$ h$_3$ }	{ h$_1$ }	{ ϕ }	{ ϕ

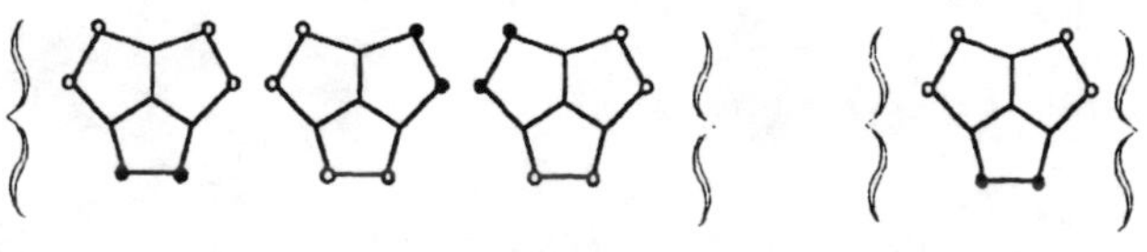

({ϕ} is an empty set). ... (208)

In eqn.(208) we draw the **actual objects** which remain fixed under the given subgroups. These "objects" model the cosets . The definition of a mark , mij, given (in its algebraic form) in section 3.23 is now much more transparent and may be well appreciated by people who prefer the non-mathematical approach! Our approach will also help with the concept of group subduction as will be shown later .

Table 35 illustrates this type of modeling of the mark tables. Indeed it models table 18 , the mark table of the $\mathbf{D_3}$ point – group.

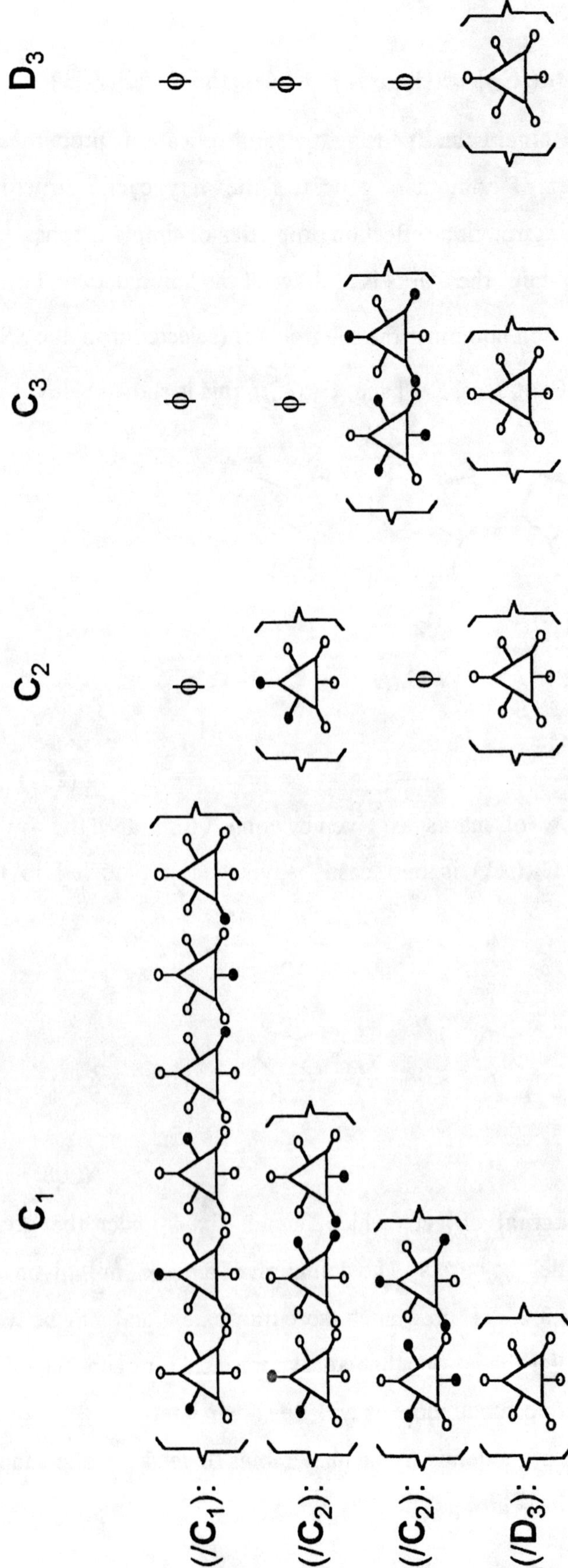

Table 35:
Modeling mark table of $\mathbf{D_3}$ point group

The set of homomers representing a given CR are equivalent [3]**:**

To demonstrate that the set of homomers which models a given CR is composed of **equivalent** colored graphs we proceed as follows. First we define an ith mapping function f_i which generate the ith homomer , h_i:

$$f_i \rightarrow h_i \quad \text{... (209)}$$

For example , $h_1 \in \mathcal{H}\,[\mathbf{C_{3v}(/C_s)}]$ is generated by the following mapping function:

label	1	2	3	4	5	6
color	w	w	b	b	w	w

... (210)

where the label (in eqn.210) refer to labeling of the vertices of the model graph given by eqn.(198) . The symbols w and b stand for white and black respectively. When the mapping defined by (210) is carried out , h_1(given by eqn.(202)) results. We now demonstrate that the homomers of a given $\mathcal{H}\,[\mathbf{G(/G_i)}]$ are indeed **equivalent** from strictly mathematical definition. We use ,as examples, h_1 and h_2 of $\mathcal{H}\,[\mathbf{C_{3v}(/C_s)}]$, (see eqn.(207) for h_1 and h_2) . The criterion of equivalence [3] is eqn.(211) , viz.,

$$f_i(V) = f_k(g(V)) \leftrightarrow h_i \sim h_k \quad \text{... (211)}$$

where g is an element of symmetry belonging to $\mathbf{C_{3v}}$ and $f_i \rightarrow h_i$ while $f_k \rightarrow h_k$ (i = 1 and k = 2 in our case since we are considering h_1 and h_2) . The mapping function f_2 , which generates h_2 is given by eqn.(212) which is to be compared with eqn.(210):

label	1	2	3	4	5	6
color	b	b	w	w	w	w

... (212)

We now apply eqn.(211) for g = C_3' , i.e. rotation by 120° in a counterclockwise way,i.e., rotation by 240° in a clockwise manner):

$f_1(1) = w$, $f_2(C_3'(1)) = f_2(5) = w$;

$f_1(2) = w$, $f_2(C_3'(2)) = f_2(6) = w$;

$f_1(3) = b$, $f_2(C_3'(3)) = f_2(1) = b$;

$f_1(4) = b$, $f_2(C_3'(4)) = f_2(2) = b$;

$f_1(5) = w$, $f_2(C_3'(5)) = f_2(3) = w$;

$f_1(6) = w$, $f_2(C_3'(6)) = f_2(4) = w$ . ... (213)

Whence $f_1 \sim f_2$ and $h_1 \sim h_2$ ▪

We observe that **the** element of symmetry which we applied in eqn.(211) is C_3', because by definition homomers are isomers which can be obtained from one another by (proper) rotation. Now h_1 is transformed into h_2 by the operation of C_3', whence our choice for g.It is "interesting" here to compare **mathematical** versus **chemical** way of definitions: For a mathematician to say that h_1 and h_2 are equivalent , he must demonstrate the equalities of a number of equations = 2 |Δ| , i.e., twice the number of vertices (12 eqns. in the present case!) , **and** to reach the desired equations , he must , of course , define the mapping functions for the configurations to be tested .

A chemist, however , would be (perfectly!) satisfied to check **just one** eqn.! viz.,

$$h_i \xrightarrow{g} h_j \quad \text{... (214)}$$

(in this case i = 1 ; j = 2 ; g = C_3')

Indeed eqn.(214) is the **chemists model** of eqn.(213) , because the rotation operation can actually be analyzed by eqns.(213). I.e., the rotation operation is "contained" in eqns.(213).The two "roads" are indeed equivalent but graphs shall always remain the more appealing for scientists when compared with algebra.

The set of homomers of a given CR is arbitrary :

From eqn.(201) it is clear that two other sets of homomers are possible depending on our selection of the color equality. However **any** of the resulting sets leads to the same permutation representations of the given coset and whence generates the same set of marks. As an illustration : if we select vertices 2 and 5 to be black we obtain the following set:

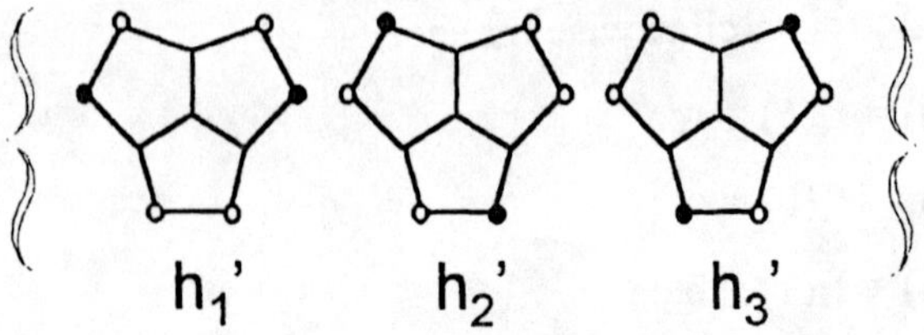

$$\{ h_1' \quad h_2' \quad h_3' \} \quad \text{... (215)}$$

The above set leads to the same marks obtained previously [c.f. eqn.(207)] . This invariance of the set of marks to one's selection of black vertices finds analogy in linear algebra [4]: An adjacency matrix depends on how one labels the corresponding graph and supposing the number of vertices in our graph is n one can construct n! adjacency matrices. However the characteristic polynomial (and whence the set of eigenvalues)of any of the n! matrices are independent of graph - labeling.

Convention

In the set $\mathcal{H}[\mathbf{G}(/\mathbf{G}_i)]$, the homomer which stays fixed under $\mathbf{G}_i$ will be called h_1.Other homomers are generated by applying all g's $\in$ **G** to h_1. In certain cases **all** the members of $\mathcal{H}$ stay fixed under $\mathbf{G}_i$, in which case we just pick one homomer and call it h_1 (see,e.g, Figs 48,52), c.f. section 5.1.3, and 5.2.5.

4.1.4 Other illustrations of the modeling method :

We now consider some other cases which further the reader's understanding of this model: (In all cases refer to **Figs** 46-54 as well as **Figs**. 1-3).

Example 39 $\mathbf{D}_{2d}(/\mathbf{S}_4)$:

Fig 57 shows the effect of all elements of $\mathbf{S}_4 = \{I, S_4, C_2, S_4'\}$ on a model graph of the $\mathbf{D}_{2d}$ point group.

Figure 57 (see page 209)

We conclude the following two sets of (color) equalities :

$$\{1 = 3 = 5 = 7\} \; ; \{2 = 4 = 6 = 8\} \qquad \ldots (216)$$

We select **either** one to be black (and the other one to be white) and arrive at the models shown in **Fig**.52

Example 40 $\mathbf{D}_{2d}(/\mathbf{D}_2)$:

Using the same labeling as in **Fig**.57 we obtain using all g $\in$ $\mathbf{D}_2$, viz., $\{I, C_2, C_2', C_2''\}$the following result :

$$\begin{matrix} C_2 \rightarrow \\ C_2' \rightarrow \\ C_2'' \rightarrow \end{matrix} \quad \begin{pmatrix} 1=5 \\ 1=8 \\ 1=4 \end{pmatrix} \begin{matrix} ; \\ ; \\ ; \end{matrix} \quad \begin{pmatrix} 2=6 \\ 2=7 \\ 2=3 \end{pmatrix} \begin{matrix} ; \\ ; \\ ; \end{matrix} \quad \begin{pmatrix} 3=7 \\ 3=6 \\ 8=5 \end{pmatrix} \begin{matrix} ; \\ ; \\ ; \end{matrix} \quad \begin{pmatrix} 4=8 \\ 4=5 \\ 6=7 \end{pmatrix} \qquad \ldots (217)$$

And using the first two columns one arrives at :

$$\{1 = 4 = 5 = 8\} \; ; \{2 = 3 = 6 = 7\} \qquad \ldots (218)$$

Indeed , the last two columns in eqn.(217) also generate the same set of vertex color equalities as in eqn.(218).When the vertices of the (sub)set to the left are colored black h_I results as shown in **Fig**.52.

Example 41 $D_{3d}(/C_{2h})$:

We operate on the model shown in **Fig**.52 with C_2 , σ_I and σ_d , i.e., the elements of C_{2h} , the result is shown in Fig.58 . One obtains the following color identities :

$$\left.\begin{array}{ll} \text{first column} \rightarrow & 1 = 2 = 9 = 10 \\ \text{third column} \rightarrow & 4 = 5 = 7 = 12 \\ \text{fourth column} \rightarrow & 3 = 6 = 8 = 11 \end{array}\right\} \quad \ldots (219)$$

Selecting the first set of equalities to be our black vertices leads to h_I of **fig**.8.

Example 42 $D_{3d}(/C_{3v})$:

We use the same model shown in **Fig**.58 and apply the symmetry operations of C_{3v} to obtain the following identities:

$$\left.\begin{array}{ll} C_3' \rightarrow & 1 = 3 = 5 ; 2 = 4 = 6 ; 7 = 9 = 11 ; 8 = 10 = 12 \\ C_3 \rightarrow & 1 = 3 = 5 ; 2 = 4 = 6 ; 7 = 9 = 11 ; 8 = 10 = 12 \\ \sigma_d \rightarrow & 1 = 10 ; 2 = 9 ; 3 = 8 ; 4 = 7 ; 5 = 12 ; 6 = 11 \\ \sigma_d' \rightarrow & 1 = 8 ; 2 = 7 ; 3 = 12 ; 4 = 11 ; 5 = 10 ; 6 = 9 \\ \sigma_d'' \rightarrow & 1 = 12 ; 2 = 11 ; 3 = 10 ; 4 = 9 ; 5 = 8 ; 6 = 7 \end{array}\right\} \quad \ldots (220)$$

Eqns. (220) generate several sets of vertex - color equalities : one such set is:

$$\{ 1 = 3 = 5 = 8 = 10 = 12 \} \quad \ldots (221)$$

Selecting the color of the above set of vertices to be white leads to h_I , while coloring them in black gives h_2 (c.f. **Fig**.52).

Figure 58 (see page 210)

Example 43 $D_{3d}(/C_2)$:

C_2 is a rather simple (sub-)group , = {I, C_2} = {I, i}, which leads to many sets of color equalities . Following the same labeling as in **Fig**.58 we obtain :

$$\{1 = 9 ; 2 = 10 ; 3 = 11 ; 4 = 12 ; 5 = 7 ; 6 = 8\} \quad \ldots (222)$$

The homomer h_1 of **Fig**.52 assigns black color to vertices 2 and 10 . Of course , other homomers are generated by operating on h_1 by all g's $\in$ $\mathbf{D}_{3d}$.

***An Important Comment !**

A reader may raise the question :

The colored configuration shown below :

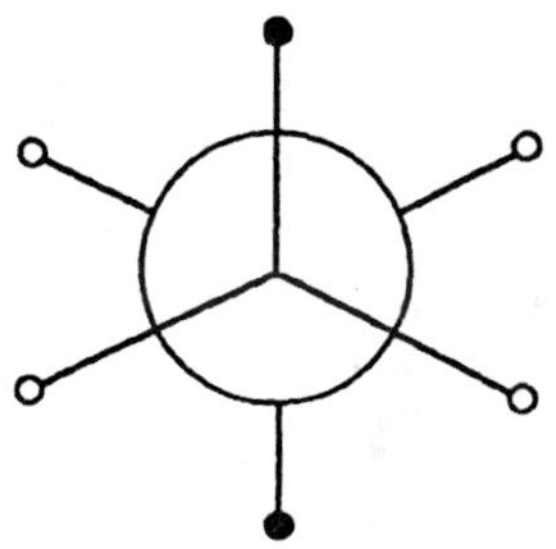

... (223)

remains invariant under $\mathbf{C}_2$.Can we ,then , use this model to obtain the marks of $\mathbf{D}_{3d}(/\mathbf{C}_2)$? The answer is **NO**! , because the model of eqn.(223) **also** remains invariant under $\mathbf{C}_s$ and $\mathbf{C}_s$' and whence the set $\mathcal{H}$ in this case leads to **wrong** marks as shown in **Fig**.59.

Figure 59 (see page 211

On the other hand the correct h_1 which remains invariant **only** under the elements of $\mathbf{C}_2$ leads to correct row of marks as portrayed in **Fig**.60 [see also **Fig**.52].

Figure 60 (see page 212)

4.2 Graphical Modeling of Subduction Tables :

Eqn.(131) is the algebraic description of the subduction of a group by one of its subgroups . This process is essential to obtain the USCI's (eqn. (132)), so that one can construct the required fixed - point and isomer - count vectors (eqns.(110,111)) . In this section we provide a convenient model for this operation . Namely the set of homomers , $\mathcal{H}[\mathbf{G}(/\mathbf{G}_j)]$ = $\{h_1,h_2,\ldots,h_r\}$ which models a given coset representation , $\mathbf{G}(/\mathbf{G}_j)$, is represented by a complete digraph on r vertices and r loops. Each vertex corresponds to a particular homomer (i.e. $h_i \leftrightarrow V_i$) so that an arc from the ith vertex ,

V_i to the jth vertex, V_j, is colored with a set S_{ij} of symmetry elements of **G** which transform h_i into h_j, i.e.,

$$\left.\begin{array}{l} g_{ij}h_i = h_j\ ; \\ g_{ij} \in \mathcal{S}_{ij} \end{array}\right\} \qquad \text{... (224)}$$

A loop associated with V_i is colored with a set S_{ii} which contains the elements of symmetry that leaves the ith homomer, h_i, fixed, i.e.

$$\left.\begin{array}{l} g_{ii}h_i = h_i\ ; \\ g_{ii} \in \mathcal{S}_{ii} \end{array}\right\} \qquad \text{... (225)}$$

Naturally g_{ii} and g_{ij} belongs to **G**.

Figs.61-66 illustrate these models for several selected CR's of some of the chemically important groups considered in this book. The idea finds analogy with the so - called **Cayley color diagrams** [5] which describe the symmetry relations among the elements of a given point - group. Indeed we may call our models Cayley color diagrams of a given CR and give it the symbol $\mathcal{D}\ [\mathcal{H}\ [\mathbf{G}(/\mathbf{G}_j)]$.

Figures 61-66 (see pages 213-220)

Example44 $\mathfrak{H}$ [$\mathbf{D_3(/C_2)}$]:

The homomers , h_1, h_2, h_3 of $\mathbf{D_3(/C_2)}$ are shown if **Fig**.48. We now investigate the effect of all g $\in$ $\mathbf{D_3}$ on each homomer ; the results are outlined below:

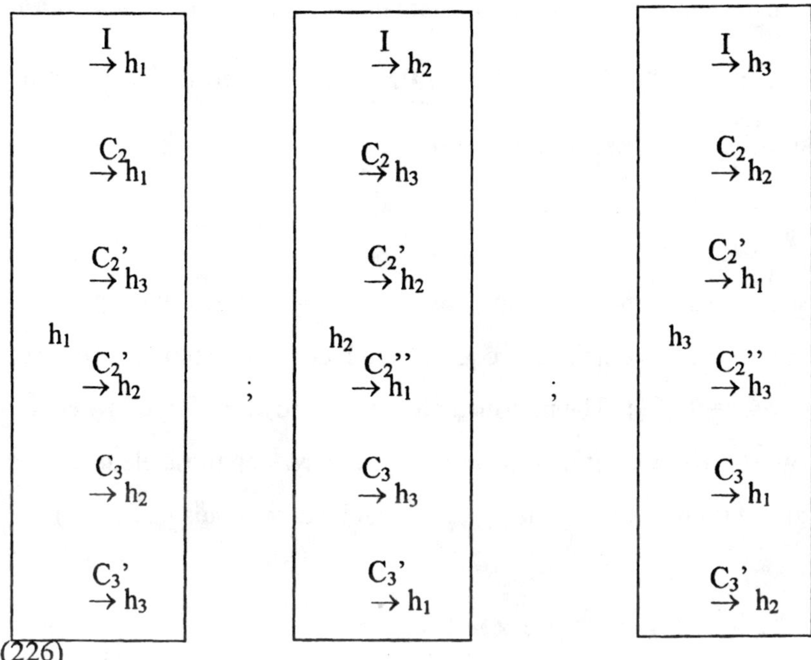

(226)

(The elements of symmetry are defined in **Fig** 1).

The above inter - relations are outlined in the Cayley diagram shown in **Fig**.62 where we replaced {I,C_2} on the loop of υ_1 (corresponding to h_1) by {$\mathbf{C_1}$,$\mathbf{C_2}$} ; the subgroups which can be generated from $\mathcal{S}_{ii}$; and similarly for the other loops. Then we can read the marks of a given CR directly off such diagrams from the color of the loops . We observe that $\mathbf{C_1}$ is repeated 3 times while $\mathbf{C_2}$ occurs only once , whence the row of marks corresponding to this CR is

$$\begin{matrix} \mathbf{C_1} & \mathbf{C_2} & \mathbf{C_3} & \mathbf{D_3} \\ (3 & 1 & 0 & 0) \end{matrix} \quad \ldots (227)$$

as reached out previously using $\mathcal{H}$ [$\mathbf{D_3(/C_2)}$] ; c.f. **Fig**.48. This (second alternative) helps one to further insights into the meaning of a mark. But this is only a "bonus" , because the main objective of these Cayley - like graphs is to model the subduction process . We provide two methods for this objective , viz.,

4.2.1 The First Method : Pruning method.

Suppose we wish to find $\mathbf{G}(/\mathbf{G}_j) \downarrow \mathbf{G}_i$ [c.f.eqn.(131)], which stands for subduction of the CR : $\mathbf{G}(/\mathbf{G}_j)$ by another subgroup , $\mathbf{G}_i$. This is equivalent to pruning out of $\mathcal{B}[\mathbf{G}(/\mathbf{G}_j)]$ all the arcs which are colored by elements of symmetry which do **not** belong to $\mathbf{G}_i$. i.e. an arc in $\mathcal{B}[\mathbf{G}(/\mathbf{G}_j)]$ is kept **only** if it belongs to $\mathbf{G}_i$.The resulting graphs are to be compared with those in figs.61-66.

***Examples**

In **Fig**.67 we illustrate this pruning method for $\mathbf{D}_2(/\mathbf{C}_1) \downarrow \mathbf{C}_2$.The Cayley graph of $\mathbf{D}_2(/\mathbf{C}_1)$ is shown in **Fig**.62. To obtain the required subduction we keep only the arcs which are colored by $\mathbf{C}_2 = \{I, C_2\}$. The intermediate step is shown in **Fig**.67 where all arcs colored with g's $\notin$ $\mathbf{C}_2$ are printed in grey. In the third step these elements are totally annihilated to end up with two Cayley graphs , each corresponds to $\mathbf{C}_2(/\mathbf{C}_1)$.

Figure 67 (see page 221)

This can be told immediately from **Fig**.61. Then we can easily see why $\mathbf{D}_2(/\mathbf{C}_1) \downarrow \mathbf{C}_2 = 2\mathbf{C}_2(/\mathbf{C}_1)$!

Figure 68 (see page 222)

In **Fig**.68 we exemplify subductions of $\mathbf{D}_2(/\mathbf{C}_2)$ by $\mathbf{C}_2$, $\mathbf{C}_2'$ and $\mathbf{C}_2''$.

Figure 69 (see page 223)

Fig.69 shows graphical derivations of $\mathbf{D}_{2d}(/\mathbf{C}_2) \downarrow \mathbf{S}_4$ and $\mathbf{D}_{2d}(/\mathbf{C}_s) \downarrow \mathbf{C}_s$.

Figure 70-71 (see page 224-225)

Fig.70 still illustrates this pruning concept using $\mathbf{C}_{2v}(/\mathbf{C}_1) \downarrow \mathbf{C}_s$ while **Fig**.71 illustrates the graphical derivation of $\mathbf{C}_{3v}(/\mathbf{C}_s) \downarrow \mathbf{C}_s$.

4.2.2 The Second Method : Build -Up Process.

An alternative approach to find the result of $\mathbf{G(/G_j)} \downarrow \mathbf{G_i}$ is to start with the set of homomers of $\mathbf{G(/G_j)}$, apply the elements of symmetry of $\mathbf{G_i}$ **only**, draw the resulting Cayley color graph, and finally identify it from the standard ones shown in figs.61-66.

Example 45 $\mathbf{D_{2d}(/C_2) \downarrow C_2}$:

The set of homomers which models the CR : $\mathbf{D_{2d}(/C_2)}$ may be found in **Fig**.52 (four homomers) We now construct a digraph based on the effect of (I,C_2) **only** on the four homomers. It is easy to concieve the following relations :

$$h_1 \begin{array}{l} \xrightarrow{I} h_1 \\ \xrightarrow{C_2} h_1 \end{array} ; \quad h_2 \begin{array}{l} \xrightarrow{I} h_2 \\ \xrightarrow{C_2} h_2 \end{array} ; \quad h_3 \begin{array}{l} \xrightarrow{I} h_3 \\ \xrightarrow{C_2} h_3 \end{array} ; \quad h_4 \begin{array}{l} \xrightarrow{I} h_4 \\ \xrightarrow{C_2} h_4 \end{array} \qquad \ldots (228)$$

The inter - relations outlined by eqn.(228) lead to four copies of $\mathfrak{G}[C_2(/C_2)]$, see **Fig**.61 , viz.,

{I,C₂} {I,C₂} {I,C₂} {I,C₂}

h₁ h₂ h₃ h₄

... (229)

Whence one writes the following algebraic description of (229):

$$\mathbf{D_{2d}(/C_2) \downarrow C_2} = 4C_2(/C_2) \qquad \ldots (230)$$

The above result may also be reached starting with the Cayley diagram of $\mathbf{D_{2d}(/C_2)}$ shown in **Fig**.65 and keeping only the arcs and loops which carry the components of $\mathbf{C_2}$ in which case the graph disintegrates into 4 "lonely" vertices whose loops are colored with the elements of $\mathbf{C_2}$.

Example 46 $\mathbf{D_{2d}(/C_2') \downarrow S_4}$:

The set which models $\mathbf{D_{2d}(/C_2')}$ is composed of 4 homomers shown in **Fig**.52.The multiplication properties of these colored graphs with respect to the elements of symmetry of $\mathbf{S_4}$ are shown below:

	I	C_2	S_4	S_4
h_1	$\to h_1$	$\to h_3$	$\to h_4$	$\to h_2$
h_2	$\to h_2$	$\to h_4$	$\to h_1$	$\to h_3$
h_3	$\to h_3$	$\to h_1$	$\to h_2$	$\to h_4$
h_4	$\to h_4$	$\to h_2$	$\to h_3$	$\to h_1$

... (231)

The above relations lead to the Cayley diagram of $\mathbf{S_4(/C_1)}$ drawn in **Fig**.66 and thence we can write :

$$\mathbf{D_{2d}(/C_2') \downarrow S_4 = S_4(/C_1)} \quad \text{... (232)}$$

We observe in all these Cayley diagrams that the vertices are left unlabeled because one can still reach the required result without the need to label the vertices.

Example 47 $\mathbf{T(/C_2) \downarrow C_3}$:

The set of (six) homomers which models $\mathbf{T(/C_2)}$ is shown in **Fig**.54. To obtain the required expression for subduction we operate on each of these homomers by the elements of symmetry of $\mathbf{C_3}$. The results are outlined below

	I	C_3	C_3'
h_1	$\to h_1$	$\to h_2$	$\to h_3$
h_2	$\to h_2$	$\to h_3$	$\to h_1$
h_3	$\to h_3$	$\to h_1$	$\to h_2$
h_4	$\to h_4$	$\to h_5$	$\to h_6$
h_5	$\to h_5$	$\to h_6$	$\to h_4$
h_6	$\to h_6$	$\to h_4$	$\to h_5$

... (233)

When the above "multiplication properties" are graphed they lead to a disconnected digraph composed of two components , each of which corresponds to the Cayley graph of $\mathbf{C_3(/C_1)}$, c.f. **Fig**.61 and therefore we write:

$$\mathbf{T(/C_2) \downarrow C_3 = 2C_3(/C_1)} \quad \text{... (234)}$$

An Important Note:

This second method is particularly suited when the graph of the unsubduced CR , ($T(/C_2)$, in this case) is too large. Indeed $\mathcal{D}(T(/C_2))$ is composed of 6 vertices , 6 loops and 30 arcs and is probably too involved for convenient hand - manipulations. For such cases we offer still another method outlined in the next section.

4.2.3 Matrix Representation of Group - Subduction:

This (third) approach still helps to envisage the algebra of the subduction operation [eqns.(131)]. However the main advantage of the matrix method over the graphical method is when the Cayley diagrams are large. In fact when the number of vertices is more than 5 the resulting digraphs become too involved for graphical manipulations. **However the advantage of our mark and subduction models remains because once one understands the simple cases , the larger ones become less opaque.**

Firstly we define a matrix which illustrates a given CR by the symbol $\mathbf{M}[G(/G_j)]$. The matrix **M** describes the multiplication properties of the elements of $\mathcal{H}[G(/G_j)]$ and therefore **M** is an r × r matrix (r = number of homomers in a given $\mathcal{H}$) . The diagonal elements of **M** are the sets of symmetry elements $\mathcal{S}_{ii}$ defined by eqn.(225) while the off - diagonal elements are the sets $\mathcal{S}_{ij}$ defined by eqn.(224).

Example 48

$\mathbf{M}(T(/C_3))$: (see **Fig**.54 for h_1 - h_4 and **Fig**.3 for the symmetry operations).

$$\begin{array}{c} \\ h_1 \\ h_2 \\ h_3 \\ h_4 \end{array} \begin{array}{cccc} h_1 & h_2 & h_3 & h_4 \\ \left(\begin{array}{c} \{I, C_{3(1)}, C_3'_{(1)}\} \\ \{C_{2(1)}, C_{3(3)}, C_3'_{(4)}\} \\ \{C_{2(2)}, C_{3(2)}, C_3'_{(3)}\} \\ \{C_{2(3)}, C_{3(4)}, C_3'_{(2)}\} \end{array}\right. & \begin{array}{c} \{C_{2(1)}, C_{3(4)}, C_3'_{(3)}\} \\ \{I, C_{3(2)}, C_3'_{(2)}\} \\ \{C_{2(3)}, C_{3(3)}, C_3'_{(1)}\} \\ \{C_{2(2)}, C_{3(1)}, C_3'_{(4)}\} \end{array} & \begin{array}{c} \{C_{2(2)}, C_{3(3)}, C_3'_{(2)}\} \\ \{C_{2(3)}, C_{3(1)}, C_3'_{(3)}\} \\ \{I, C_{3(4)}, C_3'_{(4)}\} \\ \{C_{2(1)}, C_{3(2)}, C_3'_{(1)}\} \end{array} & \left.\begin{array}{c} \{C_{2(3)}, C_{3(2)}, C_3'_{(4)}\} \\ \{C_{2(2)}, C_{3(4)}, C_3'_{(1)}\} \\ \{C_{2(1)}, C_{3(1)}, C_3'_{(2)}\} \\ \{I, C_{3(3)}, C_3'_{(3)}\} \end{array}\right) \end{array}$$

...(235)

However one does not really need to label the rows and columns with the corresponding homomer as we will see shortly.

In **Fig**.72 we show several $\mathbf{M}[G(/G_j)]$'s for several of the basic point groups where the diagonal elements list the subgroups which can be formed from the elements of $\mathcal{S}_{ii}$.

Figure 72 (see page 226-227)

This type of presentation has the advantage of listing the marks by simple inspection of the corresponding **M**. One simply counts the subgroups that occur along the main diagonal of **M**. Now the set $\mathbf{G(/G_j) \downarrow G_i}$ might be obtained by subducing $\mathbf{M[G(/G_j)]}$ according to the following steps:

i) Annihilate from $\mathbf{M[G(/G_j)]}$ all $g \notin \mathbf{G_i}$ to give $\mathbf{M[G(/G_j) \downarrow G_i]}$

ii) If $\mathbf{M[G(/G_j) \downarrow G_i]}$ is not already in block form, then apply to it the appropriate set of row/column operations [21] to transform it into the general block form :

$$\begin{pmatrix} \mathbf{M[G_i(/G_k)]} & \mathbf{0} & \mathbf{0} & \cdots & \mathbf{0} \\ \mathbf{0} & \mathbf{M[G_i(/G_\ell)]} & \mathbf{0} & \cdots & \mathbf{0} \\ \vdots & & & & \\ \mathbf{0} & \mathbf{0} & \mathbf{0} & \cdots & \mathbf{M[G_i(/G_m)]} \end{pmatrix} \quad \ldots (236)$$

where $\mathbf{G_k}, \mathbf{G_\ell}, \ldots, \mathbf{G_m}$ are subgroups of $\mathbf{G_i}$.

iii) The block matrix, (237), represents the disconnected Cayley digraphs:

$$\{ \mathcal{G}[\mathbf{G_i(/G_k)}] \,.\, [\mathbf{G_i(/G_\ell)} \ldots . \,\mathcal{G}[\mathbf{G_i(/G_m)}\} \quad \ldots (237)$$

iv) The **M** matrices along the diagonal of (236) of the $\mathcal{G}$'s in (237) are to be compared with matrices representing CR's of basic point group (**Fig**.72) or their graphical representations (**Fig**.61-66), respectively.

Example 49

$\mathbf{T(/C_3) \downarrow C_2}$

To expand the above subduction into the form of eqn.(131) we keep only those sets of elements of symmetry which belong to $\mathbf{C_2}$ and then identify the resulting block matrix . If the resulting matrix is not already in block form we apply the appropriate set of row/column operations and the resulting blocks are then identified from the matrices of the smaller groups shown in **Fig**.72. For the case at hand one obtains:

$$\begin{array}{c} \\ h_1 \\ h_2 \\ h_3 \\ h_4 \end{array} \begin{pmatrix} h_1 & h_2 & h_3 & h_4 \\ \{\ I\ \} & \{C_{2(1)}\} & \{\ \phi\ \} & \{\ \phi\ \} \\ \{C_{2(1)}\} & \{\ I\ \} & \{\ \phi\ \} & \{\ \phi\ \} \\ \{\ \phi\ \} & \{\ \phi\ \} & \{\ I\ \} & \{C_{2(1)}\} \\ \{\ \phi\ \} & \{\ \phi\ \} & \{C_{2(1)}\} & \{\ I\ \} \end{pmatrix} \qquad \text{... (238)}$$

The (nonempty) blocks are identified as $\mathbf{M[C_2(/C_1)]}$ and whence the expression $\mathbf{T(/C_3)} \downarrow \mathbf{C_2}$ is expanded as two terms of $\mathbf{C_2(/C_1)}$; i.e., we write:

$$\mathbf{T(/C_3)} \downarrow \mathbf{C_2} = 2\ \mathbf{C_2(/C_1)} \qquad \text{... (239)}$$

Naturally $\{\phi\}$in (238)) is an empty set and one might write **0** instead so we can formally perform row/column matrix operations if necessary .

Example 50

$\mathbf{D_3(/C_1)} \downarrow \mathbf{C_2} = 3\ \mathbf{C_2(/C_1)}$

The Cayley graph of the CR $\mathbf{D_3(/C_1)}$ has six vertices which represent its six homomer models , c.f. **Fig**.48.This graph is too large to construct and then apply the pruning technique to expand the required subduction . In this and similar situations the matrix representation of CR is a more convenient method.(Alternatively the build - up process , section 4.2.2 may be used) The homomers of $\mathbf{D_3(/C_1)}$ shown in **Fig**.48 transform to one another according to the following matrix:

$$\mathbf{M[D_3(/C_1)]} = \begin{array}{c} \\ h_1 \\ h_2 \\ h_3 \\ h_4 \\ h_5 \\ h_6 \end{array} \begin{pmatrix} h_1 & h_2 & h_3 & h_4 & h_5 & h_6 \\ \{\ I\ \} & \{C_2\} & \{C_2'\} & \{C_2''\} & \{C_3\} & \{C_3'\} \\ \{C_2\} & \{\ I\ \} & \{C_3'\} & \{C_3\} & \{C_2''\} & \{C_2'\} \\ \{C_2'\} & \{C_3\} & \{\ I\ \} & \{C_3'\} & \{C_2\} & \{C_2''\} \\ \{C_2''\} & \{C_3'\} & \{C_3\} & \{\ I\ \} & \{C_2'\} & \{C_2\} \\ \{C_3'\} & \{C_2''\} & \{C_2\} & \{C_2'\} & \{\ I\ \} & \{C_3\} \\ \{C_3\} & \{C_2'\} & \{C_2''\} & \{C_2\} & \{C_3'\} & \{\ I\ \} \end{pmatrix} \qquad \text{... (240)}$$

To obtain the required expression for subduction (with C_2) we keep only those elements of C_2(= $\{I, C_2\}$) in the above matrix and carry out the appropriate row/column operations as shown below:

$$\begin{array}{c} \\ h_1 \\ h_2 \\ h_3 \\ h_4 \\ h_5 \\ h_6 \end{array}\begin{array}{c} \begin{array}{cccccc} h_1 & h_2 & h_3 & h_4 & h_5 & h_6 \end{array} \\ \begin{pmatrix} I & C_2 & 0 & 0 & 0 & 0 \\ C_2 & I & 0 & 0 & 0 & 0 \\ 0 & 0 & I & 0 & C_2 & 0 \\ 0 & 0 & 0 & I & 0 & C_2 \\ 0 & 0 & C_2 & 0 & I & 0 \\ 0 & 0 & 0 & C_2 & 0 & I \end{pmatrix} \end{array} = \mathbf{M}[\mathbf{D}_3(/\mathbf{C}_1) \downarrow \mathbf{C}_2] \qquad \ldots (241)$$

$$\Big\downarrow \begin{array}{l} 1.K_{45} \\ 2.H_{35} \end{array}$$

$$\begin{pmatrix} I & C_2 & 0 & 0 & 0 & 0 \\ C_2 & I & 0 & 0 & 0 & 0 \\ 0 & 0 & I & C_2 & 0 & 0 \\ 0 & 0 & C_2 & I & 0 & 0 \\ 0 & 0 & 0 & 0 & I & C_2 \\ 0 & 0 & 0 & 0 & C_2 & I \end{pmatrix} \equiv \begin{pmatrix} \mathbf{0}_2 & \mathbf{0}_2 & C_2(/C_1) \\ \mathbf{0}_2 & C_2(/C_1) & \mathbf{0}_2 \\ C_2(/C_1) & \mathbf{0}_2 & \mathbf{0}_2 \end{pmatrix} \qquad \ldots (242)$$

where K_{ij} ≡ interchange column i and column j while H_{ij} ≡ interchange row i and row j and $\mathbf{0}_2$ is a 2 ×2 null matrix . Eqn.(242) , then corresponds to three $C_2(/C_1)$'s which is what one obtains using conventional coset algebra which requires both permutation representations **and** mark tables of the subducing groups.

4.3 A worked out enumeration using the modeling method :

Find all the structures which possess the general formula

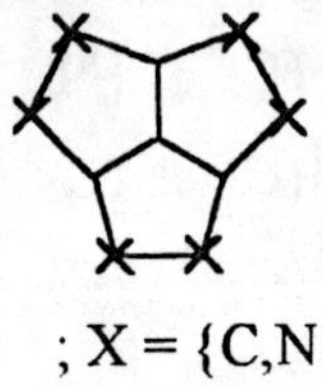

; X = {C,N} ... (243)

and classify them according to their subgroup symmetries .

Firstly we observe that this is a C_{3v} problem (see **Fig** 2 for elements of symmetry and subgroups).The model homomers are given in **Fig**.49 from which the required marks are obtained .Indeed a model of the mark **Table** is found in **Fig**.73.Here the organic chemist, the principal "beneficiary" of Fujita's work , shall see why there is a "6" followed by three zeros in the first row of the mark **Table** of $\mathbf{C_{3v}}$. This is because there are six configurations which remain fixed under $\mathbf{C_1}$ but none under any of other subgroups! He will see models of such configurations and not just silent numbers! Similarly why the second row of the mark **Table** is given as (3 1 0 0). The 3 denotes 3 configurations which are fixed by $\mathbf{C_1}$ but only **one** of them is **also** fixed by $\mathbf{C_s}$! and so on. When a reader has enough practice modeling mark tables of simple groups , he can **then** use the abstract tables found in appendices of theoretical books. The main step now is **orbit classification**. Firstly we observe that there are three orbits in this graph as shown below: (denoted by circles , squares and a triangle)

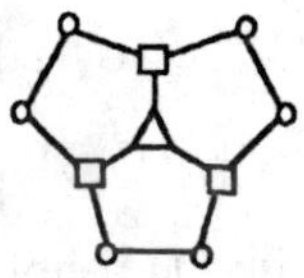

... (244)

We are concerned with the orbit which is composed of open circles. Therefore we apply the elements of symmetry to this orbit using the labeling shown in (244) and the elements of symmetry defined in **Fig**.2. The results are shown below:

	PR	$\mathbf{C_1}$	$\mathbf{C_s}$	$\mathbf{C_3}$	$\mathbf{C_{3v}}$
I	(1)(2)(3)(4)(5)(6)	✔	✔	✔	✔
C_3	(153)(264)			✔	✔
C_3'	(135)(246)			✔	✔
σ_v	(16)(25)(34)		✔		✔
σ_v'	(12)(36)(45)				✔
σ_v''	(23)(14)(56)				✔

which leads to the following row of marks:

$\mathbf{C_1}$	$\mathbf{C_s}$	$\mathbf{C_3}$	$\mathbf{C_{3v}}$
(6	0	0	0)

... (245)

Therefore this orbit is controlled by $\mathbf{C_{3v}(/C_1)}$ as can be told from the mark **Table** of $\mathbf{C_{3v}}$ shown in **Fig**.73.

Figure 73 (see page 228)

Furthermore, this is the orbit to be subduced by $\mathbf{C_1}$,$\mathbf{C_s}$,$\mathbf{C_3}$ and $\mathbf{C_{3v}}$ so that we can find the USCI's . The required subductions are computed using our matrix model because $\mathcal{G}$ [$\mathbf{C_{3v}(/C_1)}$] is too large (6 homomers ≡ 6 vertices) for pruning method . We have :

$$\mathbf{M[C_{3v}(/C_1)]} = \begin{array}{c} \\ h_1 \\ h_2 \\ h_3 \\ h_4 \\ h_5 \\ h_6 \end{array} \begin{array}{c} \begin{array}{cccccc} h_1 & h_2 & h_3 & h_4 & h_5 & h_6 \end{array} \\ \begin{pmatrix} \{I\} & \{\sigma_v'\} & \{C_3\} & \{\sigma_v''\} & \{C_3'\} & \{\sigma_v\} \\ \{\sigma_v'\} & \{I\} & \{\sigma_v''\} & \{C_3\} & \{\sigma_v\} & \{C_3'\} \\ \{C_3'\} & \{\sigma_v''\} & \{I\} & \{\sigma_v\} & \{C_3\} & \{\sigma_v'\} \\ \{\sigma_v''\} & \{C_3'\} & \{\sigma_v\} & \{I\} & \{\sigma_v'\} & \{C_3\} \\ \{C_3\} & \{\sigma_v\} & \{C_3'\} & \{\sigma_v'\} & \{I\} & \{\sigma_v''\} \\ \{\sigma_v\} & \{C_3\} & \{\sigma_v'\} & \{C_3'\} & \{\sigma_v''\} & \{I\} \end{pmatrix} \end{array} \quad \ldots(246)$$

When we keep the elements of $\mathbf{C_1}$, i.e.{I}, only in the above matrix we end up with 6 $\mathbf{C_1(/C_1)}$'s along the diagonal of (84). This is a trivial case , and we write :

$$\mathbf{C_{3v}(/C_1)} \downarrow \mathbf{C_1} = 6\mathbf{C_1(/C_1)} \quad \ldots (247)$$

and applying formulas (132) and (133) we find the corresponding USCI to be S_1^6.[because $d_{ik} = |\mathbf{C_1}|/\mathbf{C_1}| = 1$ in this case and $\beta_{ik} = 6$ c.f. eqn.132 $\therefore$ USCI = S_1^6] .

Now to obtain $\mathbf{C_{3v}(/C_1)} \downarrow \mathbf{C_s}$ we keep only $\{I, \sigma_v\}$ in $\mathbf{M[C_{3v}(/C_1)]}$ and apply the indicated row/column operations , viz.,

$$\mathbf{M[C_{3v}(/C_1)]} \xrightarrow[\mathbf{C_s}]{\text{Keep only}} \begin{array}{c} \\ 1 \\ 2 \\ 3 \\ 4 \\ 5 \\ 6 \end{array} \begin{array}{c} \begin{array}{cccccc} 1 & 2 & 3 & 4 & 5 & 6 \end{array} \\ \begin{pmatrix} I & 0 & 0 & 0 & 0 & \sigma_v \\ 0 & I & 0 & 0 & \sigma_v & 0 \\ 0 & 0 & I & \sigma_v & 0 & 0 \\ 0 & 0 & \sigma_v & I & 0 & 0 \\ 0 & \sigma_v & 0 & 0 & I & 0 \\ \sigma_v & 0 & 0 & 0 & 0 & I \end{pmatrix} \end{array} \xrightarrow[2)\ K_{15}]{1)\ H_{15}}$$

$$\begin{array}{c|cccccc} & 1 & 2 & 3 & 4 & 5 & 6 \\ \hline 1 & I & \sigma_v & 0 & 0 & 0 & 0 \\ 2 & \sigma_v & I & 0 & 0 & 0 & 0 \\ 3 & 0 & 0 & I & \sigma_v & 0 & 0 \\ 4 & 0 & 0 & \sigma_v & I & 0 & 0 \\ 5 & 0 & 0 & 0 & 0 & I & \sigma_v \\ 6 & 0 & 0 & 0 & 0 & \sigma_v & I \end{array} \equiv \begin{pmatrix} \mathbf{C_s(/C_1)} & \mathbf{0_2} & \mathbf{0_2} \\ \mathbf{0_2} & \mathbf{C_s(/C_1)} & \mathbf{0_2} \\ \mathbf{0_2} & \mathbf{0_2} & \mathbf{C_s(/C_1)} \end{pmatrix} \qquad \text{... (248)}$$

where , of course, $\mathbf{(C_s/C_1)} = \begin{pmatrix} I & \sigma_v \\ \sigma_v & I \end{pmatrix}$; $\mathbf{0_2} = \begin{pmatrix} 0 & 0 \\ 0 & 0 \end{pmatrix}$ ... (249)

[c.f.**Fig**.28].

and whence:

$$\mathbf{C_{3v}(/C_1) \downarrow C_s = 3C_s(/C_1)} \qquad \text{... (250)}$$

and because $|\mathbf{C_s}|/|\mathbf{C_1}| = 2$ the USCI associated with this subduction is $S_2^{\,3}$.

For $\mathbf{C_{3v}(/C_1) \downarrow C_3}$ we keep only $\{I, C_3, C_3'\}$ in $\mathbf{M[C_{3v}(/C_1)]}$ and perform $K_{23}, K_{35}, H_{23}, H_{35}$ to obtain

$$\begin{pmatrix} \mathbf{C_3(/C_1)} & \mathbf{0_3} \\ \mathbf{0_3} & \mathbf{C_3(/C_1)} \end{pmatrix} \qquad \text{... (251)}$$

which means that :

$$\mathbf{C_{3v}(/C_1) \downarrow C_3 = 2C_3(/C_1)} \qquad \text{... (252)}$$

corresponding to $S_3^{\,2}$

Trivially $\mathbf{C_{3v}(/C_1) \downarrow C_{3v}}$ is simply $\mathbf{C_{3v}(/C_1)}$ leading to $S_6^{\,1}$ [because $|\mathbf{C_{3v}}|/|\mathbf{C_1}| = 6$].

Here we observe that only a rudimentary **and popular** knowledge of linear algebra*[1] is used to obtain the required subductions. Now we adopt the following weights :

$$w(C) = 1 \quad ; w(N) = x \qquad \ldots (253)$$

And using theorem of Polya, we obtain the following FP matrix.

$$\begin{array}{c} \\ 1 \\ x \\ x^2 \\ x^3 \\ x^4 \\ x^5 \\ x^6 \end{array} \begin{array}{c} \begin{array}{cccc} \mathbf{C_1} & \mathbf{C_s} & \mathbf{C_3} & \mathbf{C_{3v}} \end{array} \\ \begin{pmatrix} 1 & 1 & 1 & 1 \\ 6 & 0 & 0 & 0 \\ 15 & 3 & 0 & 0 \\ 20 & 0 & 2 & 0 \\ 15 & 3 & 0 & 0 \\ 6 & 0 & 0 & 0 \\ 1 & 1 & 1 & 1 \end{pmatrix} \end{array} \qquad \ldots (254)$$

Finally from the **Table** of marks (**Fig**.73) one can almost immediately construct the required IC matrix:

$$\begin{array}{c} \\ 1 \\ x \\ x^2 \\ x^3 \\ x^4 \\ x^5 \\ x^6 \end{array} \begin{array}{c} \begin{array}{cccc} \mathbf{C_1} & \mathbf{C_s} & \mathbf{C_3} & \mathbf{C_{3v}} \end{array} \\ \begin{pmatrix} 0 & 0 & 0 & 1 \\ 1 & 0 & 0 & 0 \\ 1 & 3 & 0 & 0 \\ 3 & 0 & 1 & 0 \\ 1 & 3 & 0 & 0 \\ 1 & 0 & 0 & 0 \\ 0 & 0 & 0 & 1 \end{pmatrix} \end{array} \qquad \ldots (255)$$

The corresponding structures are shown in **Fig**.74

Figure 74 (see page 229)

* Linear algebra is usually taught at the junior or at most sophomor level in contrast to coset algebra which requires highly specialized courses.

Modeling Algebraic Properties of Mark and Group - Subduction -Tables

4.4 Properties of Marks

4.4.1 The row of marks of **G(/G$_1$)** has the general form

$$(|G| \quad O \quad O \quad \dots \quad O) \qquad \dots(256)$$

where the number of O's =|SSG|-1. This is because every loop in $\mathcal{b}$[**G(/G$_1$)**] is coloured with just one component , viz. , **C$_1$** (see **Fig.** 61-66) , and recalling that |**G(/G$_1$)**| = |**G**| = |**G[G(/G$_1$)]**| the general form of row of marks given by eqn.(256) results.

4.4.2 The row of marks of **G(/G)** has the general form :

$$(1 \quad 1 \quad \dots \quad 1) \qquad \dots(257)$$

Where the number of 1's = |SSG|. This property which one observes at the bottoms of mark **Tables** results from the fact that **G[G(/G)]** has the general form of a single vertex whose loop is coloured with every subgroup of **G**. I.e. , takes the general form:

$$\{C_1, G_2, \dots, G\} \qquad \dots(258)$$

4.4.3 The row of marks of a given **G(/G$_i$)** has only two values , viz. ,

$$m_{ij} = \begin{cases} O \\ r \end{cases} \qquad \dots(259)$$

iff all the loops of $\mathcal{b}$[**G(/G$_i$)**] are **identically** coloured with the same subgroups. Examples are shown in **Fig. 62** for **D$_2$(/C$_2$)** ; **D$_2$(/C'$_2$)** and **D$_2$(/C''$_2$)** (see **Fig. 48** e.g. for mark modeling). A counter example is the Cayley graph of **D$_{2d}$(/C$_2$')** shown in **Fig. 65** where the four loops are coloured with {C$_1$,C$_2$'} ; {C$_1$,C$_2$''}. The row of marks of **D$_{2d}$(/C$_2$')** is modeled in **Fig. 52**.

4.5 Properties of Subduced Representations

4.5.1 There are two dominant characters of the general sum given by eqn. (131) namely :

a)The number of homomers which represents (models) a given coset representation , **G(/G$_i$)** , = the number of homomers which models a given subduced representation of **G(/G$_i$)** by one of its subgroups' i.e.:

$$|\mathcal{H}[\mathbf{G}(/\mathbf{G}_i)]| = |\mathcal{H}[\mathbf{G}(/\mathbf{G}_i)\downarrow\mathbf{G}_j]| \qquad ...(260)$$

b) The two sets of homomers of eqn.(260) have identical transformation properties under all g $\in$ **G**$_j$ i.e. under all symmetry operations of the subducing groups.

These two properties may be modeled by considering for example:

$T(/C_3)\downarrow C_3 = C_3(/C_1) + C_3(/C_3)$; $C_3=\{I,C_3,C_3'\}$

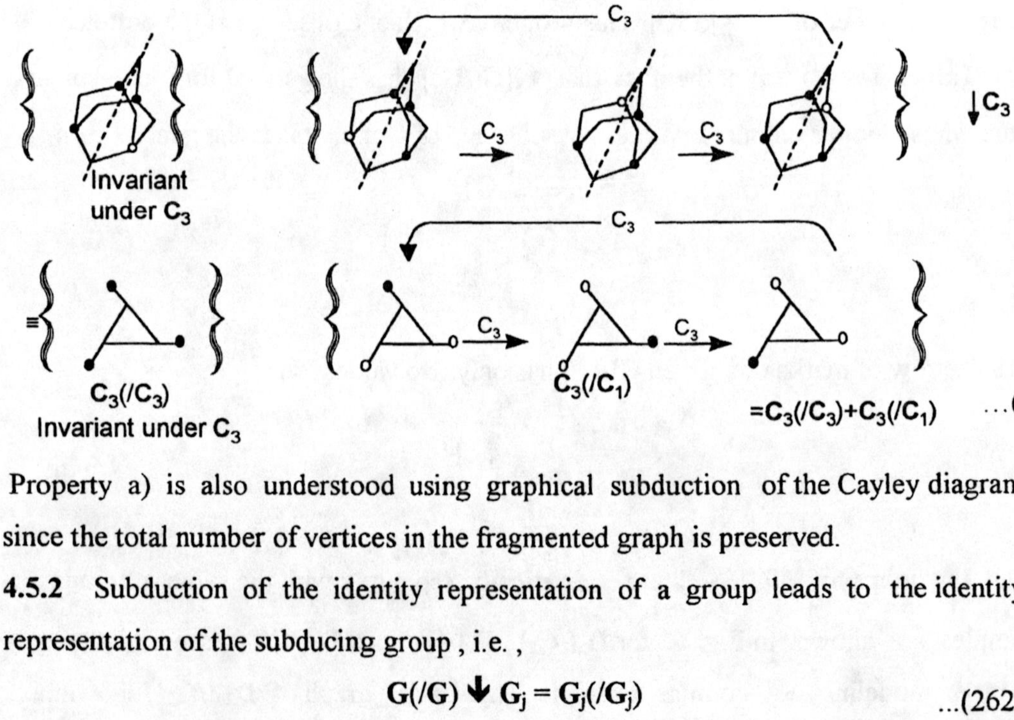

Invariant under C_3

$C_3(/C_3)$ Invariant under C_3

$C_3(/C_1)$

$=C_3(/C_3)+C_3(/C_1)$ ...(261)

Property a) is also understood using graphical subduction of the Cayley diagram since the total number of vertices in the fragmented graph is preserved.

4.5.2 Subduction of the identity representation of a group leads to the identity representation of the subducing group , i.e. ,

$$\mathbf{G}(/\mathbf{G})\downarrow \mathbf{G}_j = \mathbf{G}_j(/\mathbf{G}_j) \qquad ...(262)$$

This result is understood from the general form of the Cayley diagram of **G(/G)** , being a single vertex the loop of which is coloured with all the subgroups of **G**. Then subduction by **G**$_1$ leaves only the colour component **G**$_1$ to give **G**$_1$(/**G**$_1$) , while in general subduction by **G**$_i$ leads to a vertex whose loop is coloured by all the subgroups of **G**$_i$ which corresponds to **G**$_i$(/**G**$_i$) and so on. This property is modelled below for the identity representation of the **T** point group.

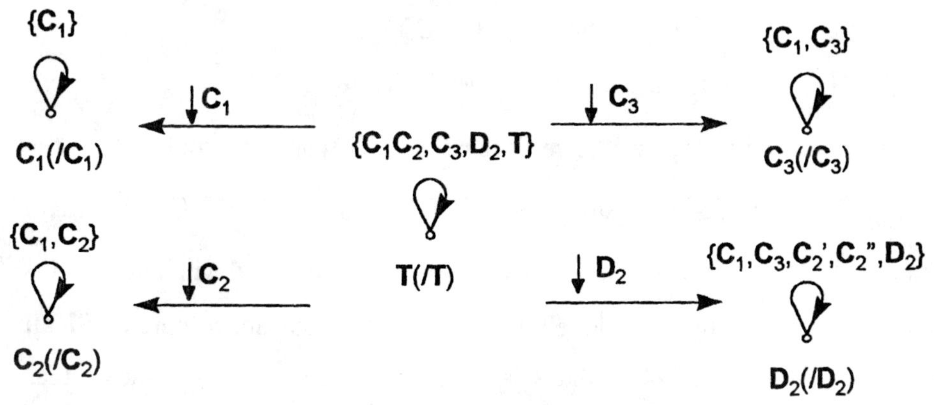

$\{C_1\}$ $C_1(/C_1)$ $\downarrow C_1$ ← $\{C_1C_2,C_3,D_2,T\}$ $T(/T)$ $\downarrow C_3$ → $\{C_1,C_3\}$ $C_3(/C_3)$

$\{C_1,C_2\}$ $C_2(/C_2)$ $\downarrow C_2$ ← $\downarrow D_2$ → $\{C_1,C_3,C_2',C_2'',D_2\}$ $D_2(/D_2)$ ...(263)

4.5.3 $$G(/G_1) \downarrow G_j = rG_j(/G_1) \quad ...(264)$$

Here the Cayley graph of $G(/G_1)$ possesses loops which are coloured by C_1 only and because C_1 is a common subgroup of all subgroups , the resulting subduced Cayley graphs will also have loops which are coloured by C_1 only and therefore they will be all identically regular representations of the subducing group , i.e. $G_j(/G_1)$. And because $r = |G| / |G_j|$, there will be r such $\mathcal{G}[G_j(/G_1)]$ generated. This property is exemplified in **Fig.** 67.

4.5.4 $$G(/G_1) \downarrow G_1 = |G|G_1(/G_1) \quad ...(265)$$

Eqn(103) is understood from the fact that the Cayley graph $\mathcal{G}[G(/G_1)]$ contains $|G|$ vertices , the loops of each of which is coloured with G_1 while the arcs with $g \notin G_1$. Then subduction by G_1 fragments it into $|G|$ vertices each of which is nothing else but $G_1(/G_1)$.Eqns (264) and (265) are modelled below:

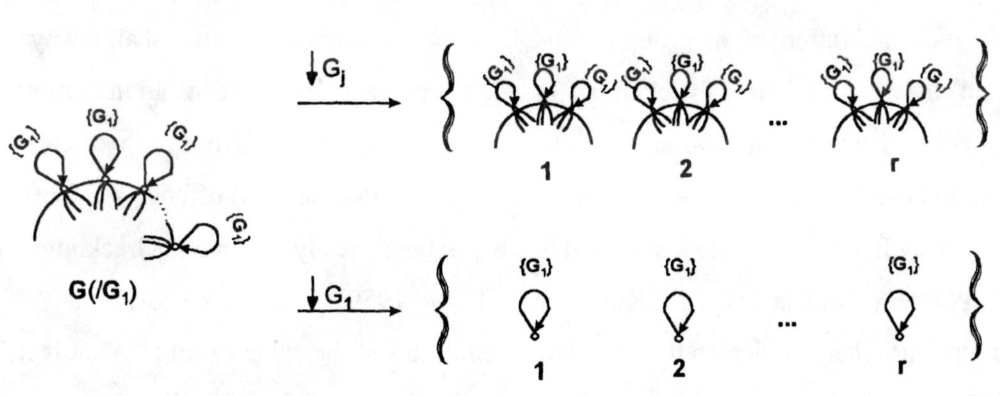

$\{G_1\}$ $G(/G_1)$ $\downarrow G_j$ → $\{ \ 1 \ 2 \ ... \ r \ \}$

$\downarrow G_1$ → $\{ \ \{G_1\}\ 1 \ \ \{G_1\}\ 2 \ ... \ \{G_1\}\ r \ \}$ ...(266)

where in the last eqn. $r = |G|/|G_1| = |G|$.

4.5.5 $$\mathbf{G}(/\mathbf{G}_i) \downarrow \mathbf{G}_i = r\,\mathbf{G}_i(/\mathbf{G}_i) \quad \ldots(267)$$

Iff all $h_i \in \mathcal{H}[\mathbf{G}(/\mathbf{G}_i)]$ remain invariant under $\mathbf{G}_i$. (c.f. **Fig.** 48 , 52)

In this particular case the loops of the vertices of $\mathcal{b}[\mathbf{G}(/\mathbf{G}_i)]$ are coloured with all the subgroups of $\mathbf{G}_i$ (because by assumption all r homomers of $\mathcal{H}[\mathbf{G}(/\mathbf{G}_i)]$ remain invariant under all $g \in \mathbf{G}_i$) and whence none of the arcs are coloured with any $g \in \mathbf{G}_i$. Then subduction by $\mathbf{G}_i$ generates r lonely vertices whose loops are coloured with all subgroups of $\mathbf{G}_i$, i.e. by $\{SSG_i\}$ which is nothing else but r copies of the regular representation of $\mathbf{G}_i$. This case is modelled below:

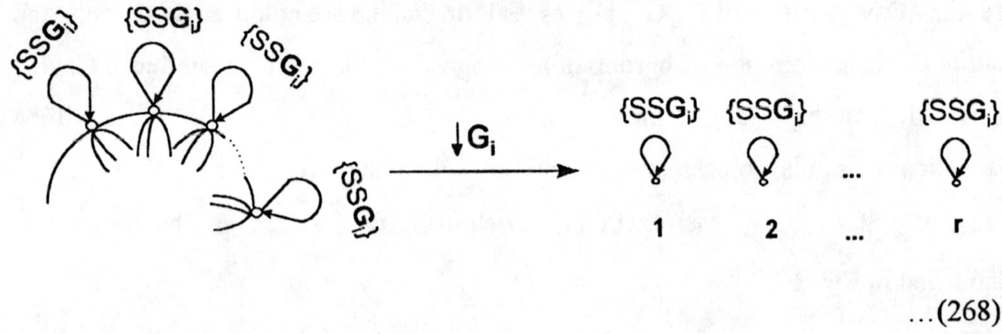

...(268)

This property is exemplified in **Fig.** 75

Figure 75 (see page 230)

* * *

In conclusion , the model presented here translates the three basic (and abstract) alphabets used in the "enumeration journey" of Fujita, viz., coset , representation , mark and subduction of a group into the language of graphs.The latter shall always remain the more appealing for chemists,Indeed the present work extends an invitation to organic chemists who would like to see organic chemistry from a theoretical (computational) vent but who are also repelled by the (sometimes) offensive algebra in their way.It may be convenient to end this paper by a parody of famous Greek myth (Rex Warner , "Men and Gods" , Kenkyusha , Tokyo 1959)[6]:

"An organic chemist demanded to know the riddle and the sphinx said : "What is it that controls elements in a group, controls atoms in a compound and finally isomers in organic chemistry ?" "Is it a coset representation or a mark ? replied the organic chemist.The sphinx found that her riddle was at last answered and died as was fated. The organic chemist received his award and he was made King of the heaven!

Bibliography

1. S. Fujita, Symmetry and combinatorial enumeration in chemistry ; Springer-Verlag : Berlin , Heidelberg and New York pp. 100,155,171.
2. R.P.Grimaldi, Discrete and combinatorial mathematics, An applied introduction, Addison-Wesley, Reading Ma., **1985** , p.85.
3. F.S.Roberts , , Applied combinatorics , Prentice-Hall , Inc.Englewood Cliffs , New Jersey , **1984** , p.299.
4. H. Anton, Elementary linear algebra , John Wiley & Sons ,New York **1981**, Chapter 1.
5. G.Chartrand, Graphs as mathematical models , Prindle , Weber & Schmidt , Incorporated , Boston , Ma , **1977**, Chapter 10 , p.232.
6. S.Fujita, Private communication, **1996**.

Chapter 5.

Application of wreath and generalized wreath products to the enumeration of stereo-position isomers.

5.1 Introduction and terminology.

A part from the scheme of Fujita[1] which depends on conjugate subgroups and mark tables of coset representations to solve discrete stereochemical problems and, in general, combinatorial enumerations in chemistry, the present chapter deals with a most suitable approach to the enumeration of stereo-position isomers, namely the wreath- and generalized-wreath product methods[2]. The approach "resolves" a molecule with a given spatial structure into a (basic) *quotient graph*, Q, and one (or more) unlabeled diagrams called *types*, T, as illustrated below in the following examples,

Example **Ethylene**

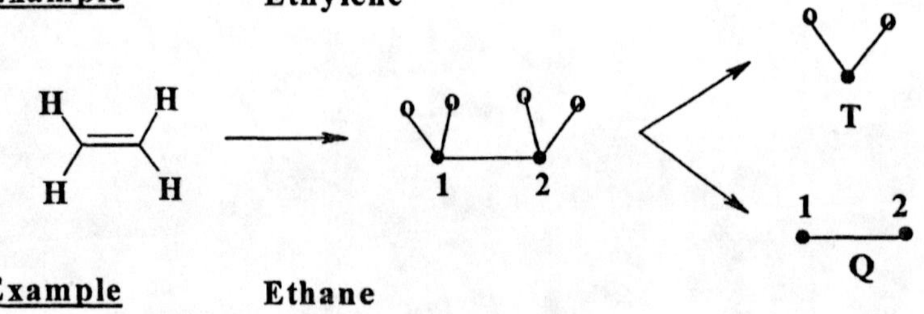

H H H H 1 2 T 1 2 Q ...(269)

Example **Ethane**

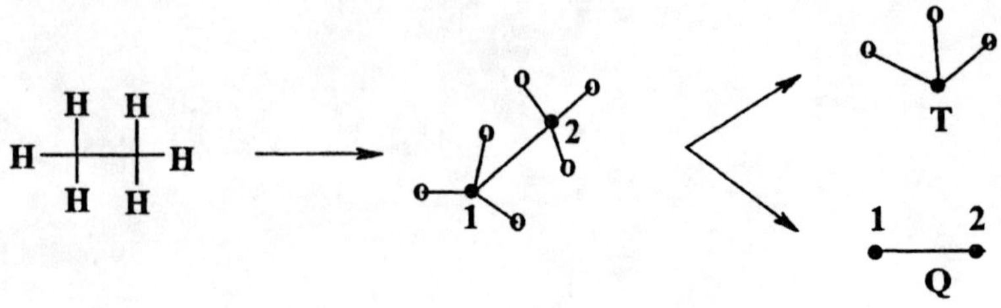

H H H H H H 2 1 T 1 2 Q ...(270)

Example **Propane**

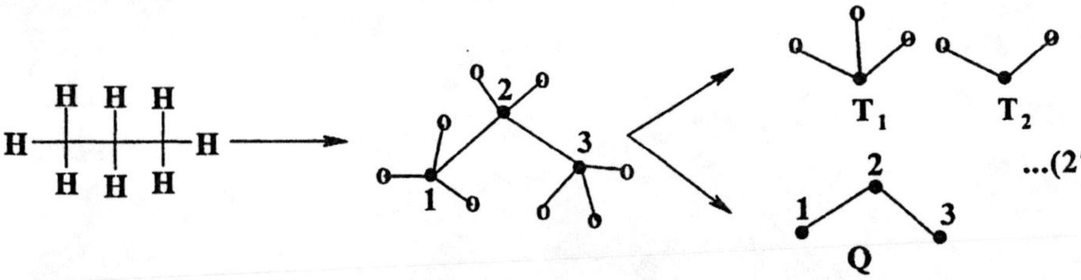

H H H H H H H H 2 3 1 T_1 T_2 2 1 3 Q ...(271)

Molecular Stereograph **Quotient and Types**

In the above examples a given molecules is drawn in a so-called *molecular stereograph*, Γ, which is defined as the graph which corresponds to the given molecule but which also includes all the chemical information regarding the spatial symmetry of the molecule. The solid (closed) vertices of Γ are called the roots are *chemically specified* (i.e. distinguishable). For example they may be all carbon atoms. The open vertices, however, are chemically unspecified. Those later vertices are where substitution is to take place. Therefore the enumeration of stereoisomers corresponds to all, non-equivalent, ways of mapping the unspecified vertices of the molecular stereograph onto a certain set of atoms or functional group. Here the definition of the wreath (and/or generalized wreath) products, (see below), allows one to control (so-to-speak) the type of rotations and/or reflections prevailing withen the Q and T parts of the Γ graph. The vertices of the Q graph are partitioned into a finite number of mutually exclusive sets according to the degrees of the vertices. For example the vertices of the Q graph of propane are partitioned into two sets, viz., $Y_1 = \{1,3\}$; $Y_2 = \{2\}$, c.f. eqn. (271). This type of partitioning allows a formal definition of the stereograph as a product between the roots of Γ and the various types. This product is called (by Balasubramanian)[2] a *molecular stereo-product* or *root-to-root product*. In the case of propane we can write,

$$\Gamma = Q\,(y_{11} \rightarrow T_{11}\,,\, y_{13} \rightarrow T_{13}\,,\, y_{22} \rightarrow T_{22}) \qquad \ldots(272)$$

In general, the element,

$y_{ir} \in Y_i$ where the first subscript indicates the label of the set (to which the element belongs) while the second subscript is a label of the element itself. Eqn. (273) might be generalized by writing

$$\Gamma = Q\,(y_{ir} \rightarrow T_{ir}) \qquad \ldots(273)$$

The symbol $y_{ir} \rightarrow T_{ir}$ denotes the root-to-root isomorphism between an element $y_{ir} \in Y_i$ and the root of a copy of the type T_i. The (isomorphic) copies of a type which is used in the construction of molecular stereo-product are called *represenatatives* of the corresponding type. The propane molecule has therefor two reprsentatives, viz., T_1 and T_2.

Our strategy is to introduce the wreath product first, then its "generalization", as Balasubramanian[2] calls it, the generaized wreath product. As with the other topics of the book we will use an intuitive, example-oriented, approach using a "modeling" device. The philosophy is that a reader knows *how* to do it with a minimum of formal description.

5.2 Geometrical Isomers

Example 51

Consider the following general formula

$$\begin{array}{ccc} R_1 & & R_2 \\ & \diagdown\!\!\!=\!\!\!\diagup & \\ R_3 & & R_4 \end{array} \qquad \ldots(274)$$

Assuming that the same compound is obtained if we interchange R_1 and R_2 or R_3 with R_4 or do both of these operations and/or turn the diagram upside down. Write the cycle index of the group of permutations and enumerate the number of structures of the above general formula.

To approach this problem we realize firstly that the equivalence classes allowed here may be called isomers. If one thinks of R_1, R_2, R_3 and R_4 as elements of the array:

$$\begin{matrix} R_1 & R_2 \\ R_3 & R_4 \end{matrix}$$

Then, the allowed permutations are the permutations of the elements in each row, and a permutation of the rows between themselves, i.e., the following 8 operations:

$$\begin{matrix} R_1\ R_2 & R_2\ R_1 & R_1\ R_2 & R_2\ R_1 \\ R_3\ R_4 & R_3\ R_4 & R_4\ R_3 & R_4\ R_3 \\ \\ R_3\ R_4 & R_3\ R_4 & R_4\ R_3 & R_4\ R_3 \\ R_1\ R_2 & R_2\ R_1 & R_1\ R_2 & R_2\ R_1 \end{matrix} \quad \ldots(275)$$

Observe that the lower row of sets of arrays is a mirror image of the upper row. For example:

$$\begin{matrix} R_1 & R_2 \\ R_3 & R_4 \\ \hline R_3 & R_4 \\ R_1 & R_2 \end{matrix} \quad \ldots(276)$$

The arrays listed in eqn. (275) are called the *wreath prouduct* group $\mathbf{S_2[S_2]}$ (read: $\mathbf{S_2}$ around $\mathbf{S_2}$ where $\mathbf{S_2} = \{(1)(2), (12)\}$). It is also called (by Pólya), "Gruppenkranz".

Now we give a formal definition of wreath prouduct and then show how to use this definition to generate the 8 operations of eqn. (275)

5.2.1 Definition of A[B]

One needs to define two groups, **A** and **B**, and two sets X and Y of objects where:

$$X = \{x_1, x_2, \ldots, x_d\} \quad \ldots(277)$$

$$Y = \{y_1, y_2, \ldots, y_e\} \quad \ldots(278)$$

While the groups **A** and **B** acts as X x Y.

|A| = number of elements in **A** = order of A, and similarly for **B**. The degree d is the number of elements in X and similarly for e. Then one may define a permutation group **A** acting on object set X where |A| is the order of **A** while |X| is its degree. Now, the composition **A[B]** (also called Gruppenkranz or wreath prouduct of "**A** around **B**") acts on X x Y. Now we let g be an element of **A** and h be an element of **B**, then for each g in A and any sequence $\{h_1, h_2, ..., h_d\}$ of d not necessarily distinct permutations in **B**, there is a unique permutation in **A[B]** written: $(g ; h_1, h_2, \ldots, h_d)$ such that for (x_i, y_i) in X x Y, we have:

$$(g ; h_1, h_2, \ldots, h_d)(x_i, y_j) = (gx_i, h_i y_j) \qquad \ldots(279)$$

Now we wonder about how many permutation **A[B]** generates , i.e what is |**A[B]**|? One may nicely conclude this order if we trace the steps leading to the definition of **A[B]**, these steps are:

1. Defne elements in **A** and elements in **B** → order of **A** and order of **B**.

2. Define elements in X and elements in Y → degree of **A** and degree of **B**.

3. For each g ∈ **A** list a sequence of d (not necessaily distinct) elements of **B** i.e. repitition is allowed.

4. Operate the expressions reached in step 3 on each x_i x y_i according to eqn. (279). Then each expression in step 3 leads to a unique permutation ∈ **A[B]**.

Suppose. **A** = $\{g_1, g_2\}$ (i.e. |**A**| = 2) and X = $\{x_1, x_2\}$ i.e., d = 2. Then we shall have the following expressions from g_1:

$\{g_1 ; h_1 h_1\}$,

$\{g_1 ; h_2 h_2\}$,

$\{g_1 ; h_1 h_2\}$,

$\{g_1 ; h_2 h_1\}$

We shall also have four expressions from g_2. Then we know that $|\mathbf{A}|$ must be a factor in $|\mathbf{A}[\mathbf{B}]|$. But we know from elementary combinatorics that there are n^r ways of arranging r elements from a set of n where order counts and repititions are allowed. Then the number of ways of arranging d not necessarily distinct elements of $\mathbf{B} = |\mathbf{B}|^d$, and whence:

$$|\mathbf{A}[\mathbf{B}]| = |\mathbf{A}|.|\mathbf{B}|^d \qquad ...(281)$$

Although the above reasoning and step-wise definition is straighforward if might leave a sense of discomfort. However modeling will make this type of composition perfectly clear:

5.2.2 Modeling A[B]:

To keep the subject in a simplest form we remain with our example for which we "hand-generated" the **B** permutations arrayed in eqn. (275). This is the $S_2[S_2]$ composition. i.e., $\mathbf{A} = S_2 = \{(1)(2), (12)\}$ and similarly for **B**. But we need to "model", so-to-speak, the two sets of elements upon which the above composition operates. A convenient model, it turns out, is a type of graphs called caterpillar trees[3] (also called Gutman trees or benzenoid trees). These objects are defined below:

5.2.2.1 Caterpillar tree[3]

The simplest way of defining a caterpillar tree, p_n $(m_1, m_2,..., m_n)$, is through the cocept of a derivative of a graph. Thus when all the end points of a graph $\mathcal{G}$ are deleted another graph $\mathcal{G}'$ results called the derivative of $\mathcal{G}$. A caterpillar tree is defined to be a tree graph (i.e. an acyclic graph) the derivative of which is a path. Then $P_n(m_1, m_2,\ldots, m_n)$ may be constructed by the addition of m_i monovalent vertices to vertex v_i

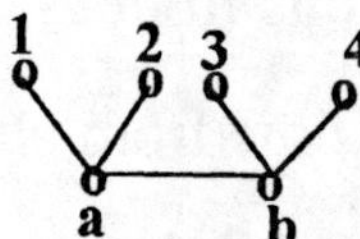

of path P_n. Then P_2 [2, 2] is the caterpillar tree shown below:

Two sets are identified, viz.,

$$X=\{a, b\} \qquad \ldots(282)$$

$$Y=\{1, 2\} \qquad \ldots(283)$$

Then $|X| = 2$; $|Y| = 2$ and $d = e = 2$. Whence we expect $|S_2[S_2]| = 2.2^2 = 8$. Further, we have:

$$\mathbf{A} = \{(a)(b), (ab)\} \; ; \qquad \ldots(284)$$

$$\mathbf{B} = \{(1)(2), (12)\} \qquad \ldots(285)$$

In this case the product X x Y is given by:

$$X \times Y = \{(a1), (a2), (b1), (b2)\} \qquad \ldots(286)$$

The cycle index of $\mathbf{S_2[S_2]}$ is obtained by the applications of eqn.(279). From the definitions of groups **A** and **B**, eqns. (284, 285), we have the following 8 operations: each operation acts on each element of (a1). The 8 operations are outlined below:

$$\left.\begin{array}{l}((a)(b)\,;(1)(2),(1)(2))\,(a1)=(a1)\\ ((a)(b)\,;(1)(2),(1)(2))\,(a2)=(a2)\\ ((a)(b)\,;(1)(2),(1)(2))\,(b1)=(b1)\\ ((a)(b)\,;(1)(2),(1)(2))\,(b2)=(b2)\end{array}\right\}\longrightarrow S_1^{\,4} \qquad \ldots(287)$$

where $S_1^{\,4}$ is obtained by the usual definition of permutation, viz.,

$$\begin{pmatrix} a1 & a2 & b1 & b2\\ a1 & a2 & b1 & b2\end{pmatrix} \sim \begin{pmatrix} 1 & 2 & 3 & 4\\ 1 & 2 & 3 & 4\end{pmatrix} \longrightarrow S_1^{\,4} \qquad \ldots(288)$$

In all cases (a)(b) is first element ($\equiv g_1$) ; (ab)$\equiv g_2$ while (1)(2) $\equiv h_1$, (12) $\equiv h_2$.

(II)

$$\left.\begin{array}{l}((a)(b)\,;(12),(12)),(a1)=(a2)\\ ((a)(b)\,;(12),(12)),(a2)=(a1)\\ ((a)(b)\,;(12),(12)),(b1)=(b2)\\ ((a)(b)\,;(12),(12)),(b2)=(b1)\end{array}\right\}\Rightarrow S_2^{\,2} \qquad \ldots(289)$$

where $S_2^{\,2}$ is obtained as usual, viz.,

$$\begin{pmatrix} a1 & a2 & b1 & b2\\ a2 & a1 & b2 & b1\end{pmatrix} \sim \begin{pmatrix} 1 & 2 & 3 & 4\\ 2 & 1 & 4 & 3\end{pmatrix}$$

(III)

$$\left.\begin{array}{l}((a)(b)\,;(1)(2),(12))\,(a1)=(a1)\\ ((a)(b)\,;(1)(2),(12))\,(a2)=(a2)\\ ((a)(b)\,;(1)(2),(12))\,(b1)=(b1)\\ ((a)(b)\,;(1)(2),(12))\,(b2)=(b2)\end{array}\right\}\Rightarrow S_1^{\,2}S_2 \qquad \ldots(290)$$

$$(g\,;h_1,h_2)\,(x_1,y_1) = (gx_1, h_1y_1)$$
$$(g\,;h_1,h_2)\,(x_1,y_2) = (gx_1, h_1y_2) \qquad \ldots(291)$$
$$(g\,;h_1,h_2)\,(x_2,y_1) = (gx_2, h_2y_1)$$
$$(g\,;h_1,h_2)\,(x_2,y_2) = (gx_2, h_2y_2)$$

i.e. the choice of h_1 or h_2 depends on the subscript of x, i.e. on j.

for example $gx_2 = (a)(b)\ b = b\,;$ ...(292)

$$h_2y_2 = (12)\ 1 = 2$$

whence the third row of eqn. (290) leads to (b2).

Observe in (292) that (a)(b) is an operator which fixes both a and b while (12) is an operator which interchanges 1 and 2.

IV

$$\left.\begin{array}{l} ((a)(b)\,;(12),(1)(2))\,(a1) = (a2) \\ ((a)(b)\,;(12),(1)(2))\,(a2) = (a1) \\ ((a)(b)\,;(12),(1)(2))\,(b1) = (b1) \\ ((a)(b)\,;(12),(1)(2))\,(b2) = (b2) \end{array}\right\} \Rightarrow S_1^2 S_2 \qquad \ldots(293)$$

Now we write four similar sets of equations but using, this time g = (ab):

V

$$\left.\begin{array}{l} ((ab)\,;(1)(2),(1)(2))\,(a1) = (b1) \\ ((ab)\,;(1)(2),(1)(2))\,(a2) = (b2) \\ ((ab)\,;(1)(2),(1)(2))\,(b1) = (a1) \\ ((ab)\,;(1)(2),(1)(2))\,(b2) = (a2) \end{array}\right\} \Rightarrow S_2^2 \qquad \ldots(294)$$

VI

$$\left.\begin{array}{l}((ab)\,;(12)\,,(12))\,(a1)=(b2)\\((ab)\,;(12)\,,(12))\,(a2)=(b1)\\((ab)\,;(12)\,,(12))\,(b1)=(a2)\\((ab)\,;(12)\,,(12))\,(b2)=(a1)\end{array}\right\}\Rightarrow S_2^{\,2} \qquad \ldots(295)$$

VII

$$\left.\begin{array}{l}((ab)\,;(1)(2)\,,(12))\,(a1)=(b1)\\((ab)\,;(1)(2)\,,(12))\,(a2)=(b2)\\((ab)\,;(1)(2)\,,(12))\,(b1)=(a2)\\((ab)\,;(1)(2)\,,(12))\,(b2)=(a1)\end{array}\right\}\Rightarrow S_4 \qquad \ldots(296)$$

VIII

$$\left.\begin{array}{l}((ab)\,;(12)\,,(1)(2))\,(a1)=(b2)\\((ab)\,;(12)\,,(1)(2))\,(a2)=(b1)\\((ab)\,;(12)\,,(1)(2))\,(b1)=(a1)\\((ab)\,;(12)\,,(1)(2))\,(b2)=(a2)\end{array}\right\}\Rightarrow S_4 \qquad \ldots(297)$$

The 8 sets of equations, (287), (289), (290), (293)-(297) lead to the required cycle index for the substituted ethylene problem, viz.,

$$Z(S_2[S_2]\,;\,S_1,\,S_2,\,S_4) = 8^{-1}[S_1^{\,4} + 3S_2^{\,2} + 2S_1^{\,2}S_2 + 2S_4] \qquad \ldots(298)$$

Using the weights r_1, r_2, r_3 and r_4 for the R_1, R_2, R_3 and R_4, respectively, one obtains the following function[4]:

$$GF = 8^{-1}\,[r_1 + r_2 + r_3 + r_4)^4 + 3(r_1^{\,2} + r_2^{\,2} + r_3^{\,2} + r_4^{\,2})^2 + 2(r_1 + r_2 + r_3 + r_4)^2(r_1^{\,2} + r_2^{\,2} + r_3^{\,2} + r_4^{\,2}) + 2(r_1^{\,4} + r_2^{\,4} + r_3^{\,4} + r_4^{\,4})] \quad ...(299)$$

The required number of isomers is then the coefficient of $r_1\ r_2\ r_3\ r_4$ in the above expansion, which is

$$8^{-1}\begin{pmatrix} 4 \\ 1\ 1\ 1\ 1 \end{pmatrix} = 3 \text{ geometrical isomers} \quad ...(300)$$

The three isomers are

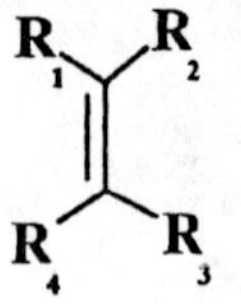

R_1, R_2 / R_4, R_3 — **1** R_1, R_3 / R_4, R_2 — **2** R_2, R_3 / R_4, R_1 — **3**

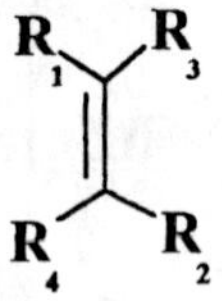

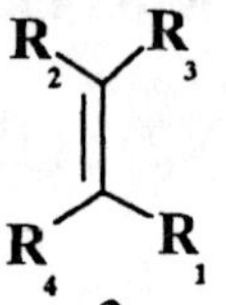

Of course structures such as

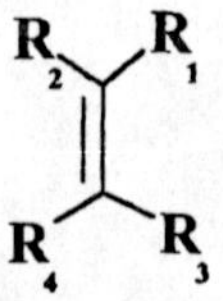

R_2, R_1 / R_4, R_3

are cnsidered identical with **1** because we assumed at the start of the problem that the same structure is obtained if we interchange R_1 and R_2 or R_3 with R_4 or do both of these operations and/or turns the diagram upside down.

5.2.3. Another Method to obtain the Cycle Index of A[B]:

An alternative approach to find the cycle index of the composition **A[B]** is to "model" a given composition as a quotient graph, Q and a type graph, T_i Namely for our example, $S_2[S_2]$ we have:

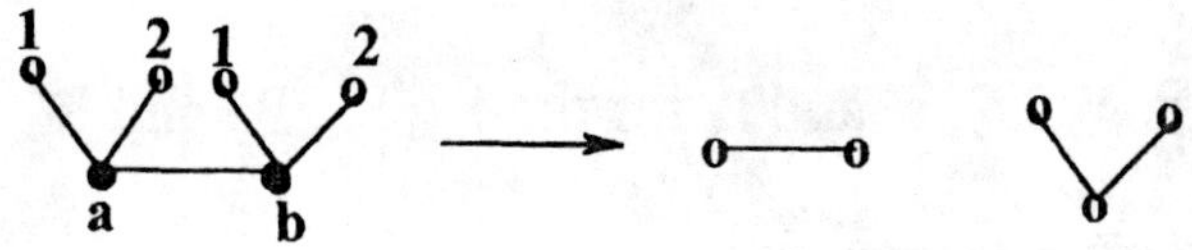

1 2 1 2 a b → Q T

The following "recipe" is used to find the required expression:

(a) Find cycle index of the quotient graph:

$$Z(Q\ ;\ S_1, S_2) = 2^{-1}[S^2{}_1 + S_2] \quad ...(301)$$

(b) Use the following substitution:

$$Z(S_2[S_2]\ ;\ S_1, S_2) = Z(Q\ ;\ S_i \rightarrow Z_i) \quad ...(302)$$

Where:

$$Z_i = Z(Q\ ;\ S_j \rightarrow S_{ij}) \quad ...(303)$$

Then:

$$Z_1 = 2^{-1}[S^2{}_1 + S_2]\ ; \quad ...(304)$$

$$Z_2 = 2^{-1}[S^2{}_2 + S_4] \quad ...(305)$$

(c) Now use (304) and (305) in (302):

$$Z(Q\ ;\ S_1S_2) = Z(Q\ ;\ S_1 \rightarrow Z_1, S_2 \rightarrow Z_2)$$

$$=2^{-1}[\{2^{-1}[S_1{}^1 + S_2]\}^2 + \{2^{-1}[S_2{}^2 + S_4]\}] \quad ...(306)$$

$$=8^{-1}[S_1{}^4 + 3S_2{}^2 + 2S_1{}^2 S_2 + 2S_4] \quad ...(307)$$

which is the same expression obtained in eqn. (298)

5.3 Stereoisomers

Example 52

The graph $C_2H_2Br_2Cl_2$ in three-dimentional Newman projection and saw-hourse convention are shown below

Newman **Saw-hourse**

The solid vertices are spcified (i.e. labeled) as carbon atoms but the other (open) vertices are unspecified. The problem amounts to finding out how many distinct ways of mapping a set which contains 2H, 2Cl and 2Br atoms onto the six open (monovalent) vertices of the stereograph shown above (in two representations). There is also an implicit information about the spatial symmetries of the molecule, viz.,

(a) A rotation of the whole molecule around the mid-point of C-C bond, i.e. permutation of the C atoms. This is the $\mathbf{C_2}$ subgroup,

(b) A rather local symmetry at each end carbon defined by a rotation of the three atoms around the C atoms; generating a $\mathbf{C_3}$ sub-group. The two subsymmeties lead to the composition $\mathbf{C_2[C_3]}$, which is modeled as shown below:

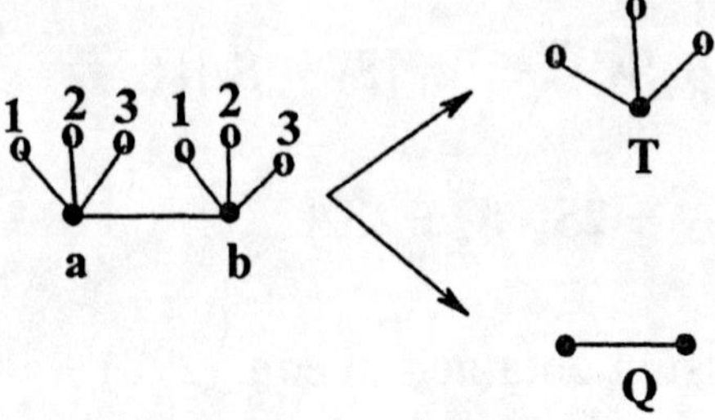

Where:

$$X = \{a, b\}\ ;\ Y = \{1, 2, 3\}\ ; \quad \dots (308)$$

$$\mathbf{C_2} = \{(a)(b), (ab)\}\ ; \quad \dots (309)$$

$$\mathbf{C_3} = \{(1)(2)(3), (123), (123)\}\ ; \quad \dots (310)$$

$$d = 2 \; ; \; e = 3 \qquad \text{...(311)}$$

Then,

$$|\mathbf{C}_2\,[\mathbf{C}_3]| = 2.3^2 = 18 \qquad \text{...(312)}$$

The following cycle indices are calculated:

$$Z\,(Q\,;\,S_1,\,S_2) = ½\,[S_1^{\,2} + S_2]\,; \qquad \text{...(313)}$$

$$Z\,(T\,;\,S_1,\,S_2,\,S_3) = 1/3\,[S_1^{\,3} + 2S_3]\,; \qquad \text{...(314)}$$

From which we derive:

$$Z_1 = 1/3\,[S_1^{\,3} + S_3]\,;\; Z_2 = 1/3\,[S_2^{\,3} + 2S_6] \qquad \text{...(315)}$$

[c.f. eqn. (303)]

We substitute eqn. (315) into eqn. (313) to obtain:

$$Z\,[\mathbf{C}_2[\mathbf{C}_3]\,;\,S] = Z\,[Q\,;\,S_i \rightarrow Z_i]$$

$$= ½\,[\{1/3[S_1^{\,3} + S_3]\}^2 + \{1/3\,[S_2^{\,3} + 2S_6]\}]$$

$$= 18^{-1}\,[S_1^{\,6} + 4S_1^{\,3}S_3 + 4S_3^{\,2} + 3S_2^{\,3} + 6S_6] \qquad \text{...(316)}$$

Now, if we use the symbols H, Br and Cl to represent the weights of the respective atoms then the required number of isomers is the coefficient of $H^2Br^2Cl^2$ in the folowing generating function:

$$GF = 18^{-1}\,[(H + Br + Cl)^6 + 4(H + Br + Cl)^3\,(H^3 + Br^3 + Cl^3) + 4(H^3 + Br^3 + Cl^3)^2 + 3(H^2 + Br^2 + Cl^2)^3 + 6(H^6 + Br^6 + Cl^6)] \qquad \text{...(317)}$$

Only the first and the fourth terms of the above inventory contain non vanishing number of $H^2Br^2Cl^2$ term, namely:

$$(H + Br + Cl)^6 \Rightarrow \frac{6!}{2!\;2!\;2!} = 90\,H^2Br^2Cl^2$$

$$3(H^2 + Br^2 + Cl^2)^3 \Rightarrow 3 \cdot \frac{3!}{1!\,1!\,1!} = 18\, H^2Br^2Cl^2$$

Therefore, there are $18^{-1}[90+18] = 6$ stereoisomers:

These are shown below:

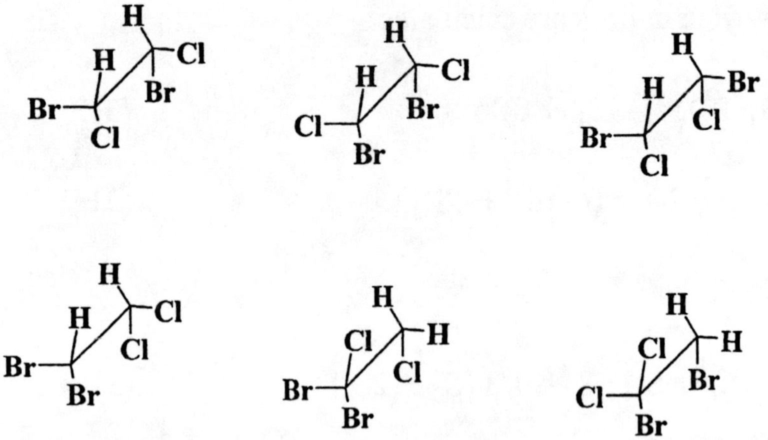

...(318)

5.4 Generalized wreath product method.

The method which will be described now is used when there is more than one type and that is why Balasabramanian[2] calls it generalized wreath-product method. It is best explained through a broblem:

5.4.1 Example 53

Comsider the folowing propane conformer

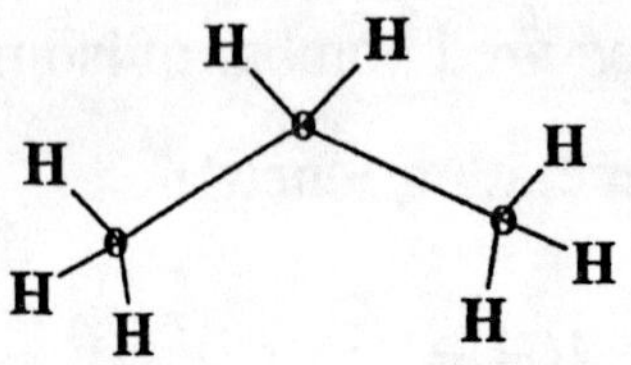

How many (distinct) mapings are there that transfer two Cl atoms to the (root) carbon atoms?

The problem here amounts to the calculation ot the number of isomers of $C_3H_6X_2$.

Analysis of the problem:

The stereograph of propane is composed of a quotient graph, Q and two types, T_1 and T_2, viz.,

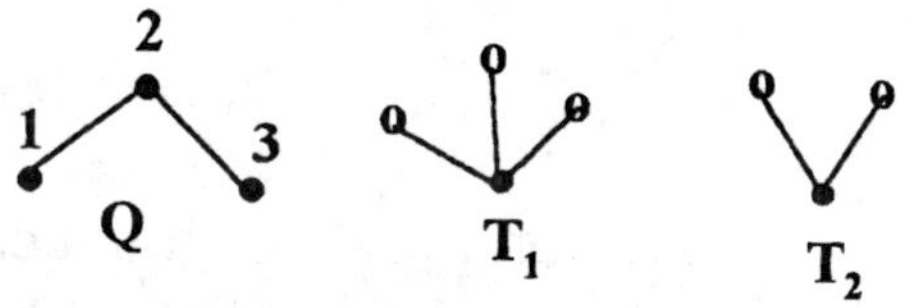

2 1 3 Q T_1 T_2

Here, Q has two sets of vertices, namely

$$Y_1 = \{1, 3\} \; ; Y_2 = \{2\}$$

Where Y_1 contains primary carbons and Y_2 a secondary carbon atom. To enumerate the stereoposition isomers at room temperature, we look for the graph **G** of all proper rotations corresponding to the stereograph Q.

At room temperature, the carbon-carbon internal rotations permute the chemically unspecified vertices of the primary carbon atoms. However, they do not permute the chemically unspecified vertices of the secondary carbon atom. This last statement makes the two H atoms of this secondary carbon nonequivalent! In our counting scheme, therefore, the two structures shown below are considered distinct, i.e. counted as two:

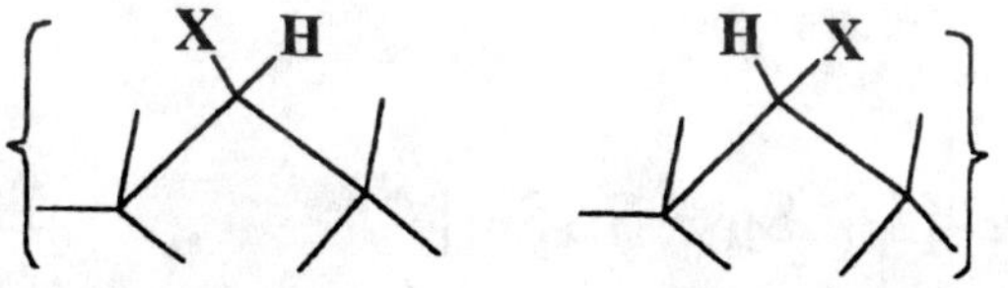

X H H X

Then the group acting on two sets of terminal H atoms is a $\mathbf{C_3}$ while that acting on the two middle H atom is the identity group, $\mathbf{C_1}$. This situation is portrayed below

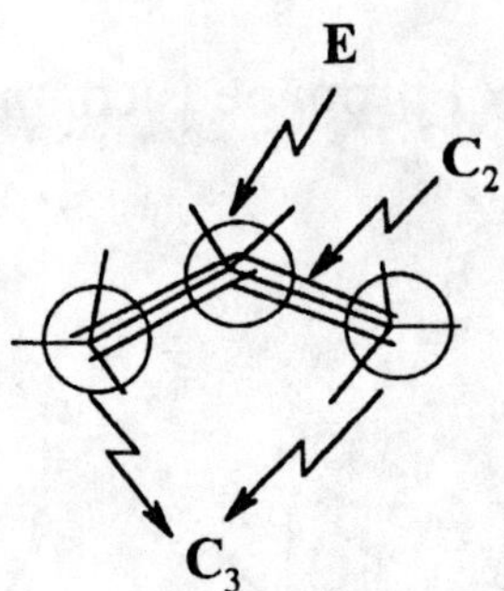

Then the "composition" acting on the stereograph can be "generalized" as:

$$\mathbf{H = C_2[C_3, C_1]} \qquad \text{...(319)}$$

Balasubramanian[2] introduced these symbols sometime ago in his study of stereo-position isomers and nmr spectra and calls them: *generalized-wreath products*.

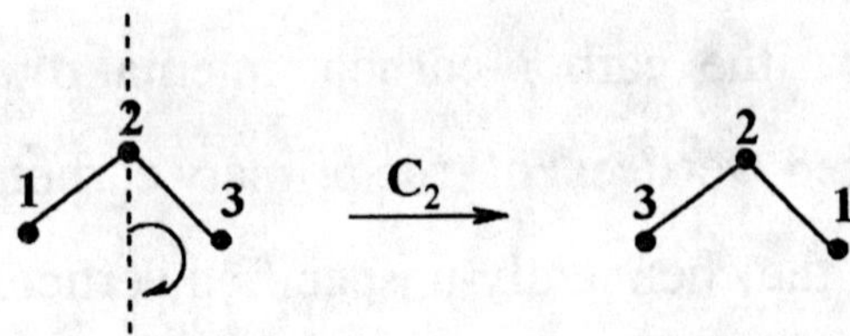

In the present case the $\mathbf{C_2}$ subgroup is illustrated for the Q graph as follows:

Therefore:

$$Z(Q) = 2^{-1}[S_{11}^{2}\, S_{21} + S_{12}\, S'_{21}] \qquad \text{...(320)}$$

Where the first subscript indicates the set (Y_1 or Y_2 in this case) to which the vertex belongs while the second subscript the cycle-length (as usual). In eqn. (320), the first term corresponds to the identity operation while the second to C_2 operation (remember that $\mathbf{C_2} = \{\mathbf{C_1}, \mathbf{C_2}\} \equiv \{I, C_2\}$). One

Now the two types lead to the following cycle indices:

T_1

$$Z_1 = 1/3\ [S_1^3 + 2S_3] \qquad \text{...(321)}$$

From which we can write two equations; viz.,

$$Z_{11} = 1/3\ [S_1^3 + 2S_3] \qquad \text{...(322)}$$
$$Z_{12} = 1/3\ [S_2^3 + 2S_6] \qquad \text{...(323)}$$

Where, in general, Balasubramanian defined:

$$Z_{ij} = Z_i\ (S_k \rightarrow S_{kj}) \qquad \text{...(324)}$$

(c.f. eqn. 303)

T_2:

$$Z_2 = S_1^2 \quad \text{(from } I\text{)} \qquad \text{...(325)}$$

Then $$Z_{21} = S_1^2 \qquad \text{...(326)}$$

$$Z'_2 = S_2 \quad \text{(from } C_2\text{)} \qquad \text{...(327)}$$

Then $$Z'_{21} = S_2 \qquad \text{...(328)}$$

Finally, we calculate the required Z(H) by appling what we may call, substituion of Balasubramanian, viz.,

$$Z(H) = Z\ (Q : S_{ij} \rightarrow Z_{ij}) \qquad \text{...(329)}$$

c.f. eqn. (302)

At this point it may be illuminating to observe that equations (302) and (329) are "isomorphic" and both are similar to eqn. (237). The three types of equations are collected below for purpose of (useful) comparison:

$\Gamma = Q\ (y_{ir} \rightarrow T_{ir})$: stereograph generation
$Z(A[B]) = Z(Q\ ;\ S_i \rightarrow Z_i)$: One type ≡ wreath product
$Z(A[B, C, \ldots]) = Z(Q\ ;\ S_{ij} \rightarrow Z_{ij})$: Several types ≡ generalized wreath product ...(330)

Eqn. (329) leads to the required cycle index, viz.,

$$Z(H) = \tfrac{1}{2}\,[\{1/3(S_1^{\,3} + 2S_3)\}^2 S_1^{\,2} + 1/3(S_2^{\,3} + 2S_6)S_2]$$

$$= 1/18[S_1^{\,8} + 4S_1^{\,5}S_3 + 4S_1^{\,2}S_3^{\,2} + 3S_2^{\,4} + 6S_6S_2] \qquad \ldots(331)$$

Where we substituted eqns. (321), (323) and (326), (328) into eqn. (320).

It may be helpful to outline the steps in the calculation of the (generalized) wreath product as follows:

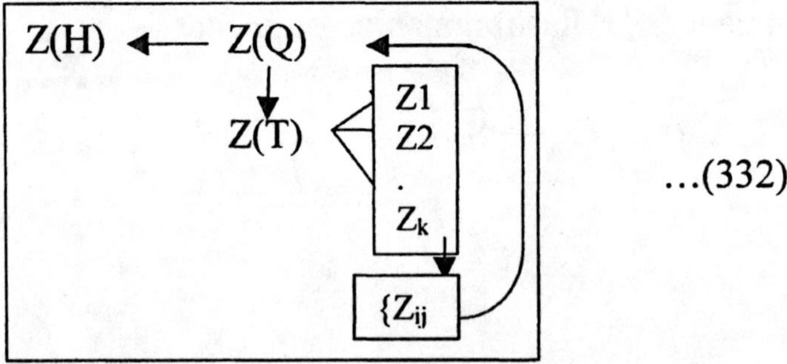

...(332)

Now to calculate the number of $C_3H_6X_2$ isomers we let the weight of H, $W(H) = \alpha_1$ while $W(X) = \alpha_2$, then the generating function is

$$\begin{aligned} GF = 18^{-1}\,[&(\alpha_1 + \alpha_2)^8 + 4(\alpha_1 + \alpha_2)^5\,(\alpha_1^{\,3} + \alpha_2^{\,3}) \\ &+ 4(\alpha_1 + \alpha_2)^2\,(\alpha_1^{\,3} + \alpha_2^{\,3})^2 \\ &+ 3(\alpha_1^{\,2} + \alpha_2^{\,2})^4 \\ &+ 6(\alpha_1^{\,6} + \alpha_2^{\,6})\,(\alpha_1^{\,2} + \alpha_2^{\,2})] \end{aligned} \qquad \ldots(333)$$

The number of stereo-position isomers is the coefficient of $\alpha_1^{\,6}\,\alpha_2^{\,2}$ which is:

$$1/18\left[\binom{8}{2} + 4\binom{5}{2} + 4\binom{2}{2} + 3\binom{4}{1} + 6\binom{1}{1}\right] = 5 \qquad \ldots(334)$$

The five isomers are drawn below: (isomers 2 and 3 are optically active).

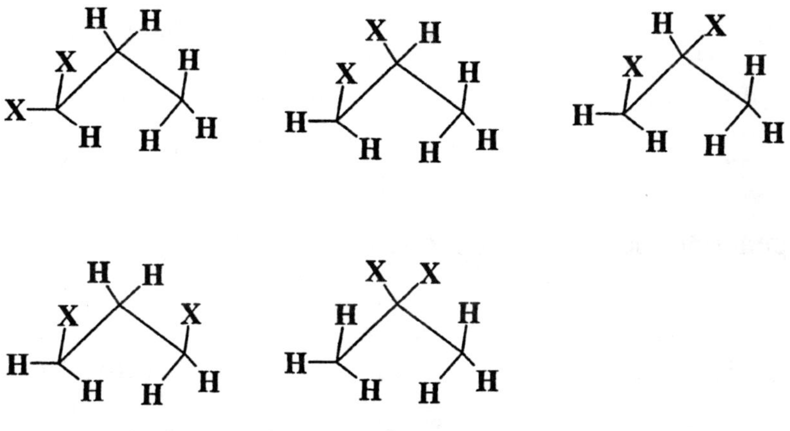

...(335)

5.4.2 Example 54

Find the number of stereo-position isomers which are generated when two H atoms of n-butane are replaced by two X atoms. i.e., find the number of stereo-position isomeric dichlorobutanes which posses the formula $C_4H_8X_2$.

***

Here there are two types, T1 and T2 and a quotient graph Q shown below together with their (parent) stereograph Γ:

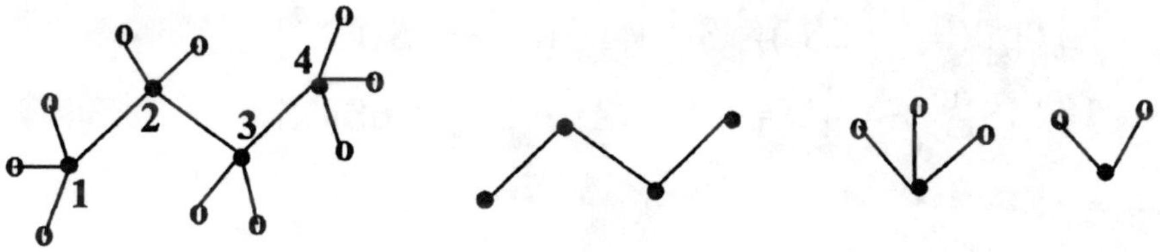

...(336)

At room temperature the Q graph is subject to $\mathbf{C_2}$ subgroup (= $\{I, C_2\}$) while the three methyl H atoms to $\mathbf{C_3}$ (= $\{I, C_3, C_3^2\}$). The model leaves both methelene hydrogens invarient. Whence we look for the cycle index of the generalised wreath product $\mathbf{C_2[C_3, C_1]}$. As in the previous example we find the following parameters:

$$Y_1 = \{1, 4\} \; ; Y_2 = \{2, 3\} \qquad ...(337)$$

$$Z(Q) = ½ [S^2_{11} S^2_{21} + S_{12} S_{22}] \quad ...(338)$$

(When the length of C-chain is even the C2-axis passes through the (2-3)-edge and whence there is no need to prime the dummy variable of the second term).

Now T_1 and T_2 generate the following cyclic structures:

T_1:

$$Z_1 = 1/3 [S_1^3 + 2S_3] \quad ...(339)$$

whence:

$$Z_{11} = 1/3 [S_1^3 + 2S_3] ; \quad ...(340)$$

$$Z_{12} = 1/3 [S_2^3 + 2S_6] \quad ...(341)$$

T_2:

$$Z_{21} = S_1^2 \quad ...(342)$$

whence:

$$Z_{21} = S_1^2 ; \quad ...(343)$$

$$Z_{22} = S_2^2 \quad ...(344)$$

Then

$$\begin{aligned} Z(H) &= Z(Q ; S_{ij} \to Z_{ij}) \\ &= ½[\{1/3(S_1^3 + 2S_3)\}^2 S_1^4 + 1/3(S_2^3 + 2S_6) S_2^2] \\ &= 18^{-1} [S_1^{10} + 4S_1^7 S_3 + 4S_1^4 S_3^2 + 3S_2^5 + 6S_2^3 S_6] \quad ...(345) \end{aligned}$$

Now we let:

$$W(H) = a ; W(X) = b \quad ...(346)$$

Then the required number of $C_4H_8X_2$ stereo-position isomers is the cosffecient of a^8b^2 in the following GF:

$$GF = 18^{-1} [(a + b)^{10} + 4(a + b)^7(a^3 + b^3) + 4(a + b)^4(a^3 + b^3)^2 + 3(a^2 + b^2)^5 + 6(a^2 + b^2)^2(a^6 + b^6)] \quad ...(347)$$

The required coefficient is calculated as follows:

* $(a+b)^{10} \rightarrow \binom{10}{2\ 8} = 45\ a^8b^2$

* $4(a+b)^7(a^3+b^3) \rightarrow$ here we look for how many a^5b^2 result from $(a+b)^7$: there are $\binom{7}{2\ 5} = 21$ such terms and whence we obtain 21x4 = <u>84</u> a^8b^2

* $4(a+b)^4(a^3+b^3)^2 = 4(a+b)^4(a^6+b^6+2a^3b^3) \rightarrow$ and we look for the number of a^2b^2 terms from $(a+b)^4$: this number is $\binom{4}{2\ 2}$ =6 and thus we obtain 6x4 = 24 a^8b^2

* $6(a^2+b^2)^5 \equiv 3(A+B)^5 \rightarrow 3 \binom{5}{4\ 1} A^4B = 15\ a^8b^2$

* $6(a^2+b^2)^2(a^6+b^6) = 6(a^4+b^4+2a^2b^2)(a^6+b^6) \rightarrow 2x6 = 12\ a^8b^2$

Then we have a total of $\frac{[45+84+24+15+12]}{18}$ =10 stereo-position isomers. These are shown below:

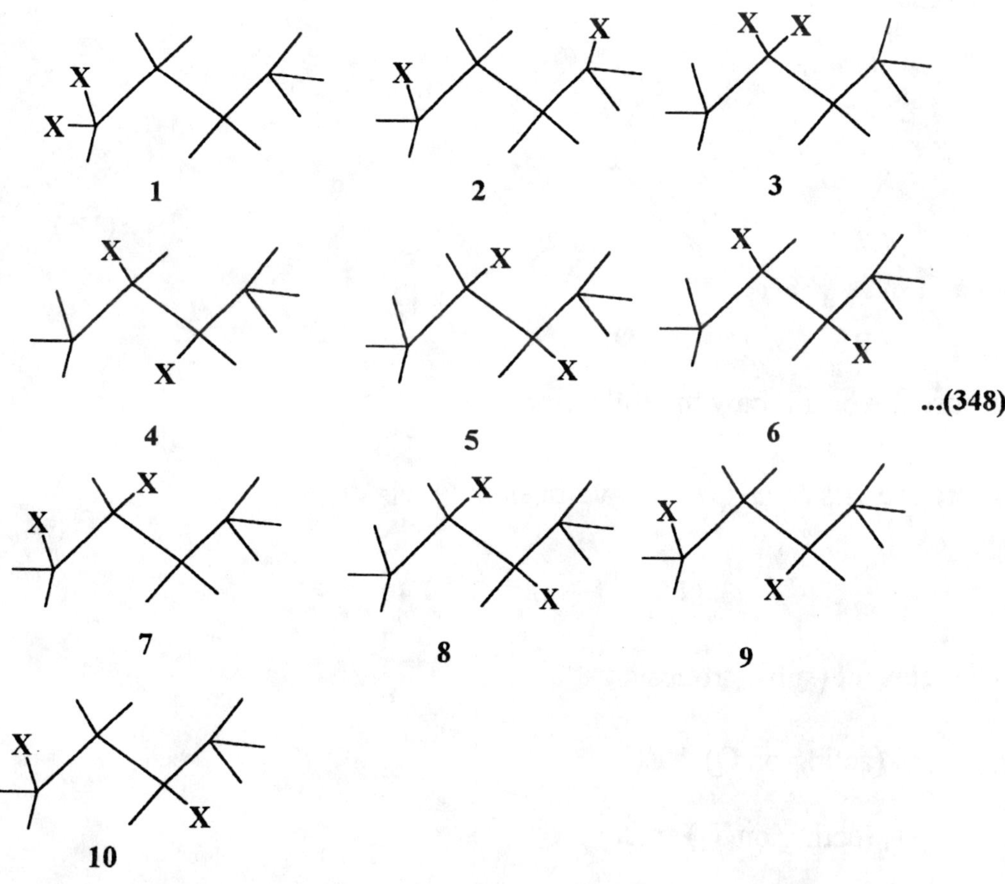

...(348)

The last 4 isomers are optically active.

Observe that when 6 is subjected to C2 (which this model assumes) it leads to 6′ and thus they are counted as one:

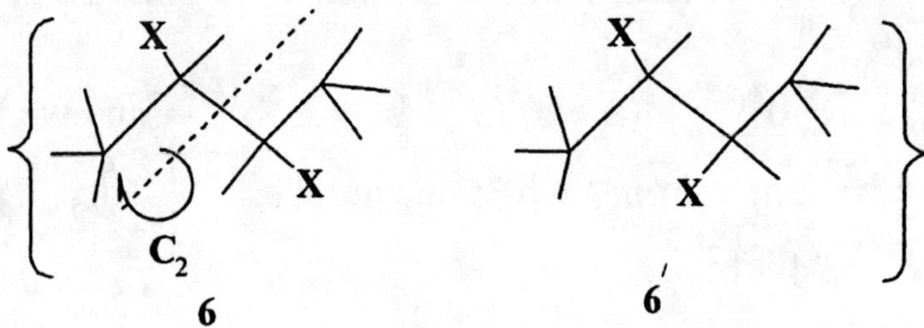

5.5 NMR using generalised wreath product.

5.5.1 Example 55

Compose a generalised wreath product for CH_3 $CHCl$ CH_2Cl and find its zero order NMR spectrum, i.e. find all equivalence calsses of the H atoms.

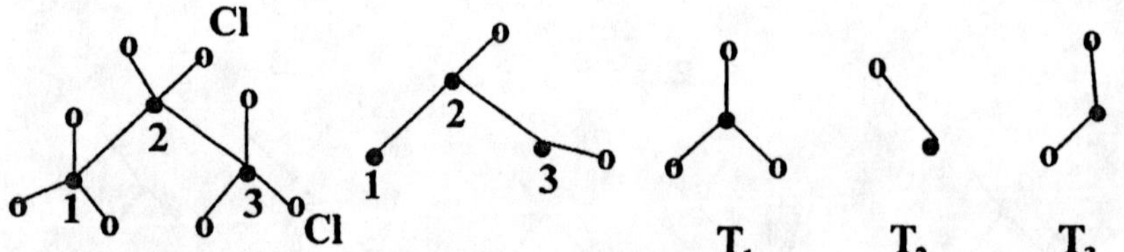

Cl 2 1 3 Cl Γ — 2 1 3 Q — T_1 — T_2 — T_3 ...(349)

As usual we can draw the following species

There are sets of unspecified vertices in Q, viz.,

$$Y_1 = \{1\} \; ; \; Y_2 = \{2\} \; ; \; Y_3 = \{3\}$$

The relevant (sub-) groups are

G (acting on Q) = $\mathbf{E_5}$

$\mathbf{H_1}$ (acting on T_1) = $\mathbf{C_3}$

$\mathbf{H}_2$ (acting on T_2) = $\mathbf{E}_1$

$\mathbf{H}_3$ (acting on T_3) = $\mathbf{E}_2$

Where $\mathbf{E}_n$ is identity operation on n atoms.

Then the generalized composition is:

$$\mathbf{H} = \mathbf{E}_5[\mathbf{C}_3, \mathbf{E}_1, \mathbf{E}_2] \quad ...(351)$$

We have

$$Z(Q) = S_{11}\, S_{21}\, S_{31} \quad ...(352)$$

$$Z_1 = 1/3[S_1^{\,3} + 2S_3] \quad ...(353)$$

$$Z_{11} = 1/3[S_1^{\,3} + 2S_3] \quad ...(354)$$

$$Z_2 = S_1 \quad ...(355)$$

$$Z_{21} = S_1 \quad ...(356)$$

$$Z_3 = S_1^{\,2} \quad ...(357)$$

$$Z_{31} = S_1^{\,2} \quad ...(358)$$

Then

$$Z(\mathbf{H}) = 1/3[S_1^{\,3} + 2S_3]S_1\, S_1^{\,2}$$
$$= 1/3[S_1^{\,6} + 2S_3\, S_1^{\,2}] \quad ...(359)$$

The inventory of NMR signals (i.e. the GF) is given by

$$GF = 1/3[(\alpha_1 + \alpha_2)^6 + 2(\alpha_1^{\,3} + \alpha_2^{\,3})\,(\alpha_1 + \alpha_2)^3] \quad ...(360)$$

The number of equivalence classes of H atoms is the coeifficient of $\alpha_1^{\,5}\, \alpha_2$ which is

$$1/3[\binom{6}{5\ \ 1} + 2\binom{3}{2\ \ 1}] = 4 \text{ classes} \quad ...(361)$$

Bibliography

1. S. Fujita, Symmetry and combinatorial enumeration in chemistry, Springer-Verlag, Berlin (1991).

2. K. Balasubramanian, A generarlized wreath product method for the enumeration of stereo and position isomers of polysubstituted organic compounds, Theor. Chim. Acta, **51,** 37-54 (1979).

3. S. El-Basil, Caterpillar (Gutman) trees in chemical graph theory, Topics in current chemistry, Springer-Verlag, Berlin, **153,** 273-289 (1990).

4. D. I. A. Cohen, Basic techniques of combinatorial theory, John Wiley & Sons, New York, Chapter 6 (1978).

Figures

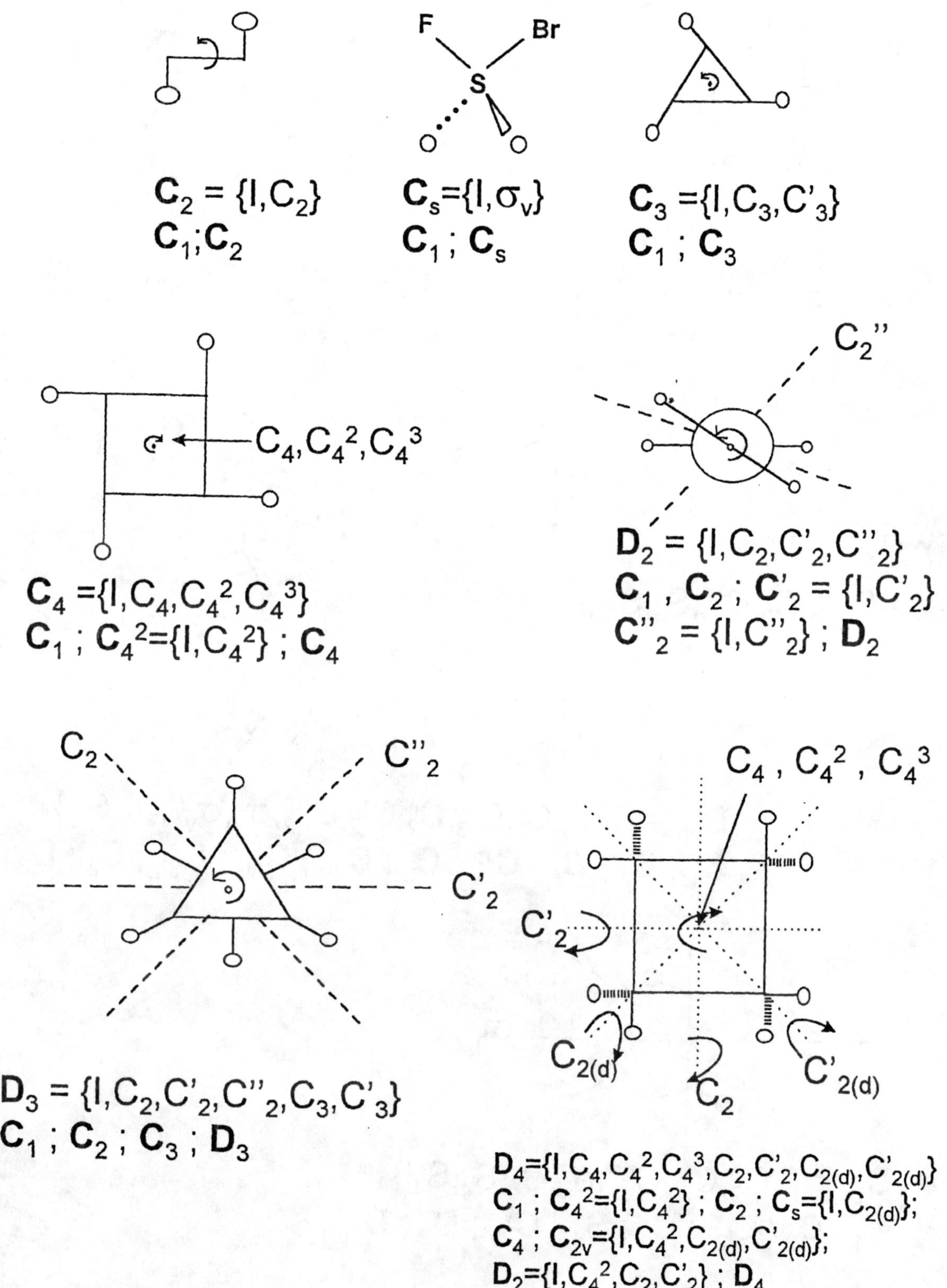

FIG1 C_2, C_3, C_4, D_2, D_3, D_4 and C_s point-group Symmetry elements and subgroup are indicated

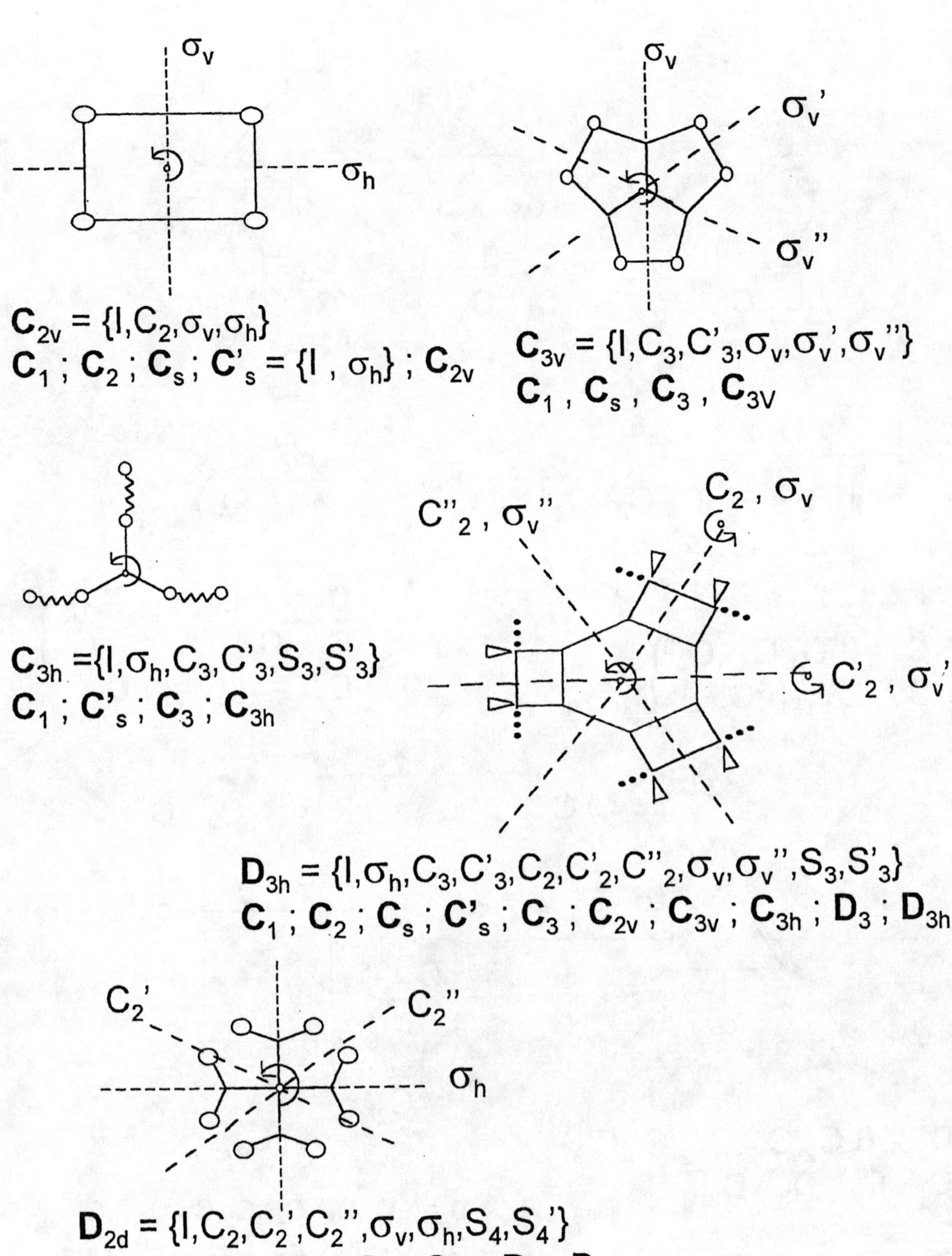

FIG2 C_{2v} , C_{3v}, C_{3h}, D_{3h}, D_{2d}, D_{3d}, and S_4 point-groups , their subgroups and their symmetry elements

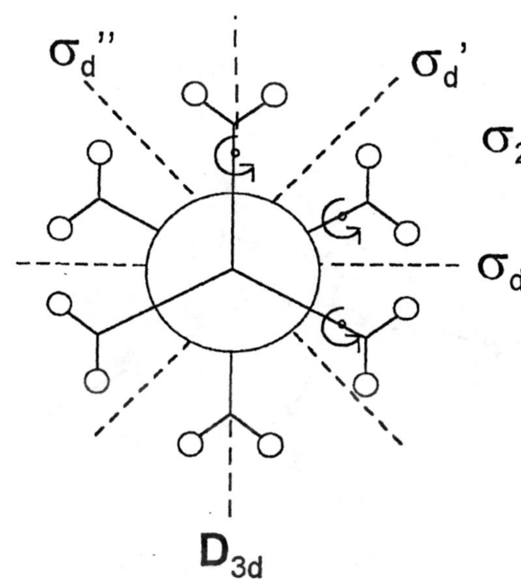

Me

Me

Me

$\mathbf{S}_4 = \{I, S_4, C_2, S'_4\}$

$\mathbf{C}_1$; $\mathbf{C}_2$; $\mathbf{S}_4$

$\mathbf{D}_{3d} = \{ I, S_6, S'_6, C_3, C_3', i(=C_2), \sigma_1, \sigma_2, \sigma_3, \sigma_d, \sigma_d', \sigma_d'' \}$

$\mathbf{C}_1$; $\mathbf{C}_s = \{I, \sigma_d\}$; $\mathbf{C}_s'\{I, \sigma_1\}$; $\mathbf{C}_2 = \{I, C_2\}$; $\mathbf{C}_3$;

$\mathbf{C}_{2h} = \{I, C_2, \sigma_1, \sigma_d\}$; $\mathbf{D}_3 = \{ I, C_3, C_3', \sigma_1, \sigma_2, \sigma_3\}$;

$\mathbf{C}_{3v} = \{I, C_3, C_3', \sigma_d, \sigma_d', \sigma_d''\}$

$\mathbf{S}_6 = \{I, C_3, C'_3, S_2, S_6, S'_6\}$; $\mathbf{D}_{3d}$

FIG 2 cont.

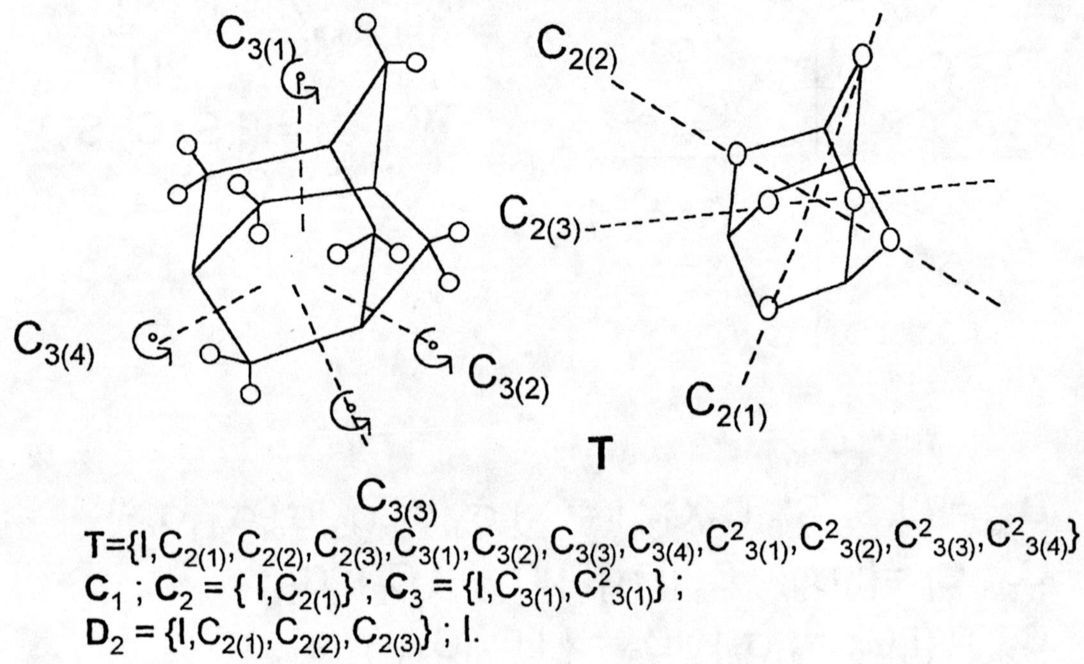

FIG3 The cubic group **T** modeled by the molecular graph of adamantane . Subgroups of **T** and its symmetry elements are indicated.

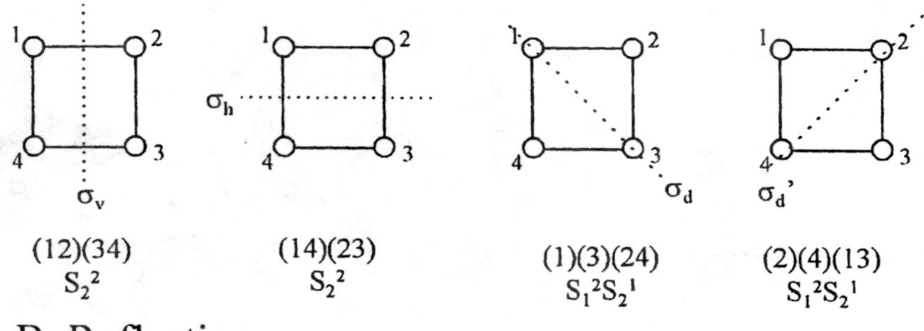

A. Rotations

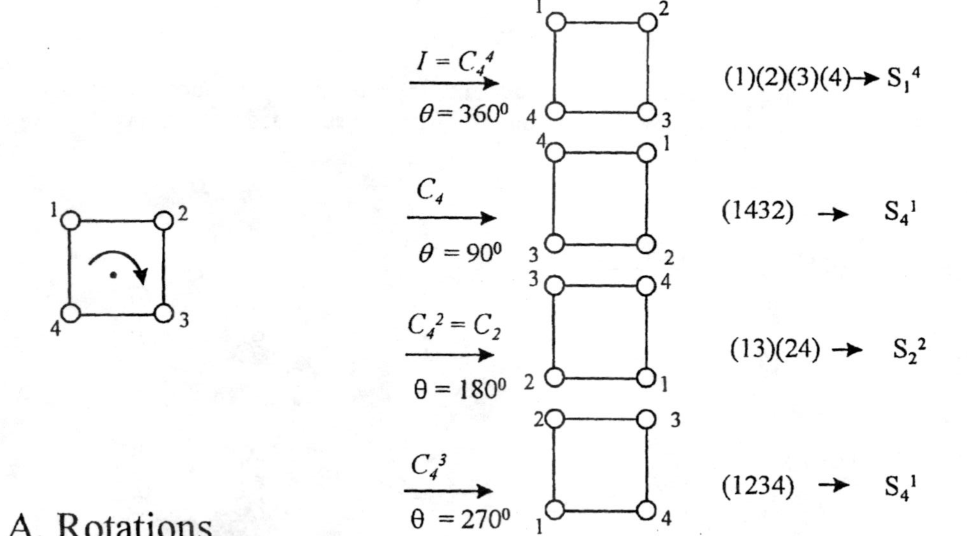

B. Reflections

FIG4 The four rotations and four reflections of the square . The symbol S_i^j indicates j i cycles.The 8 operations constitute the elements of the $\mathbf{D_4}$ point group shown in **FIG1**

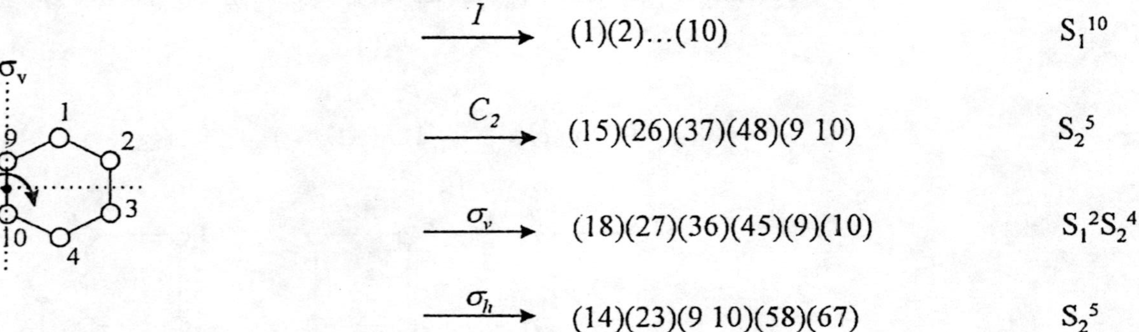

FIG5 The Symmetry operations of the molecular graph of naphthalene and the associated permutations. The four operations define the $\mathbf{C_{2v}}$ point group, c.f. **FIG2**.

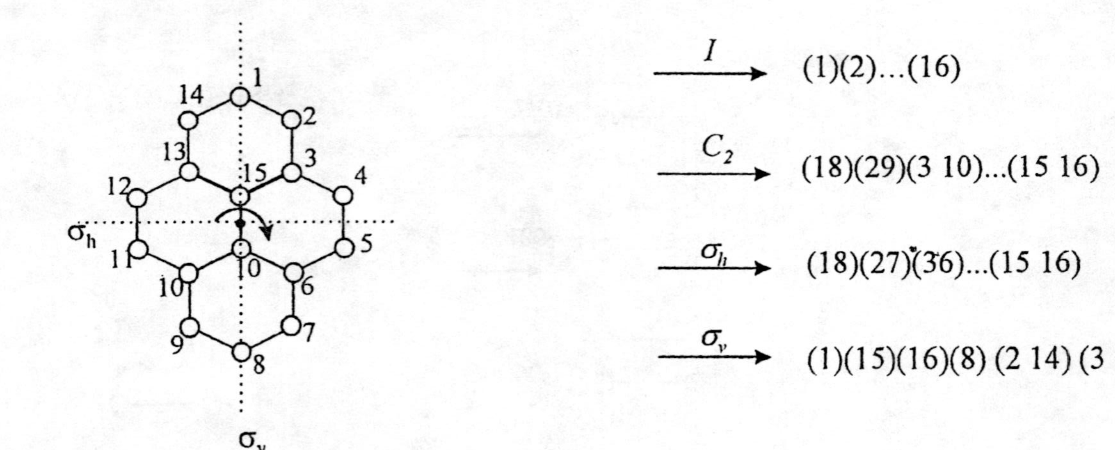

FIG6 The Symmetry operations of the pyrene molecular graph and the corresponding cycle-notations.This system belongs to the $\mathbf{C_{2v}}$ point group, c.f. **FIG2**.

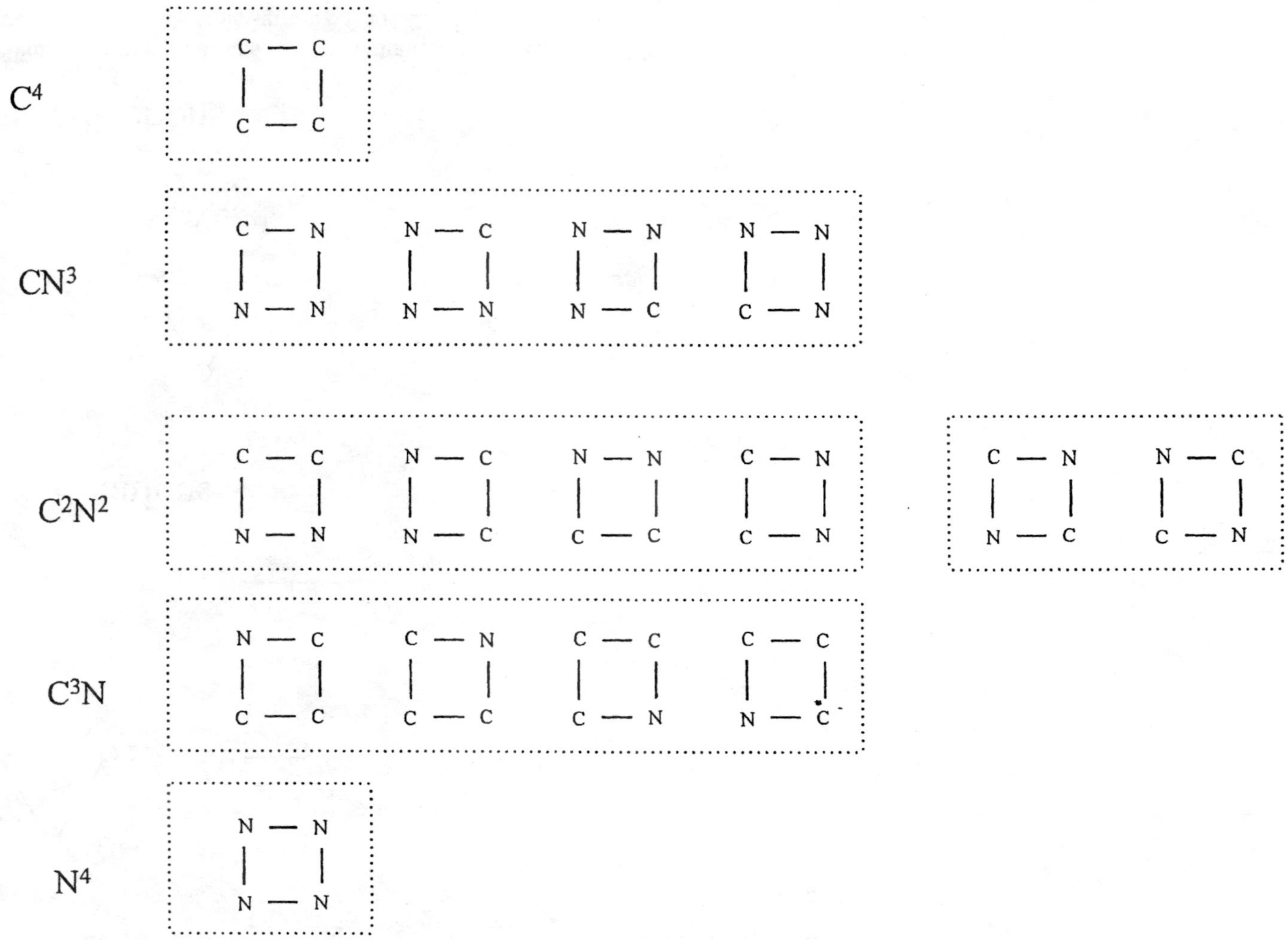

```
C4      C — C
        |   |
        C — C

CN3     C — N    N — C    N — N    N — N
        |   |    |   |    |   |    |   |
        N — N    N — N    N — C    C — N

C2N2    C — C    N — C    N — N    C — N        C — N    N — C
        |   |    |   |    |   |    |   |        |   |    |   |
        N — N    N — C    C — C    C — N        N — C    C — N

C3N     N — C    C — N    C — C    C — C
        |   |    |   |    |   |    |   |
        C — C    C — C    C — N    N — C

N4      N — N
        |   |
        N — N
```

FIG7 The six equivalence classes consistent with the general system shown in eqn.(12). Each equivalence class are transformed to one another by symmetry operation.

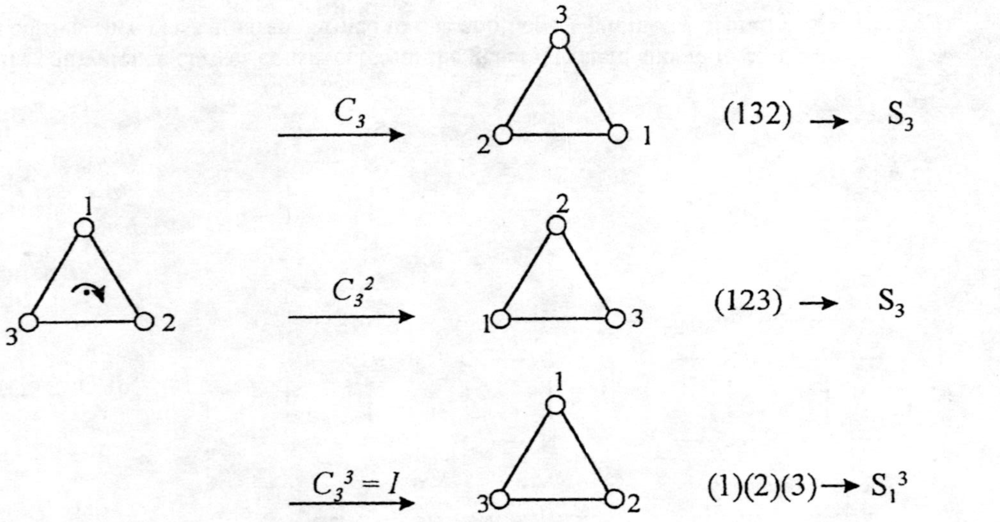

A. Rotations

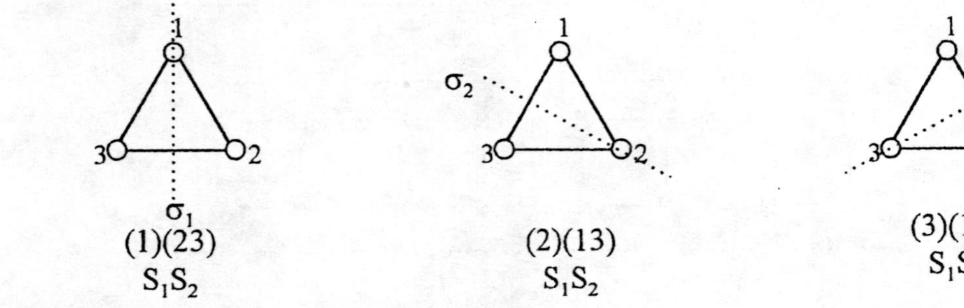

B. Reflections

FIG8 Symmetry operations and their permutations of the equilateral triangle , an example of the C_{3v} point-group c.f. **FIG2**.

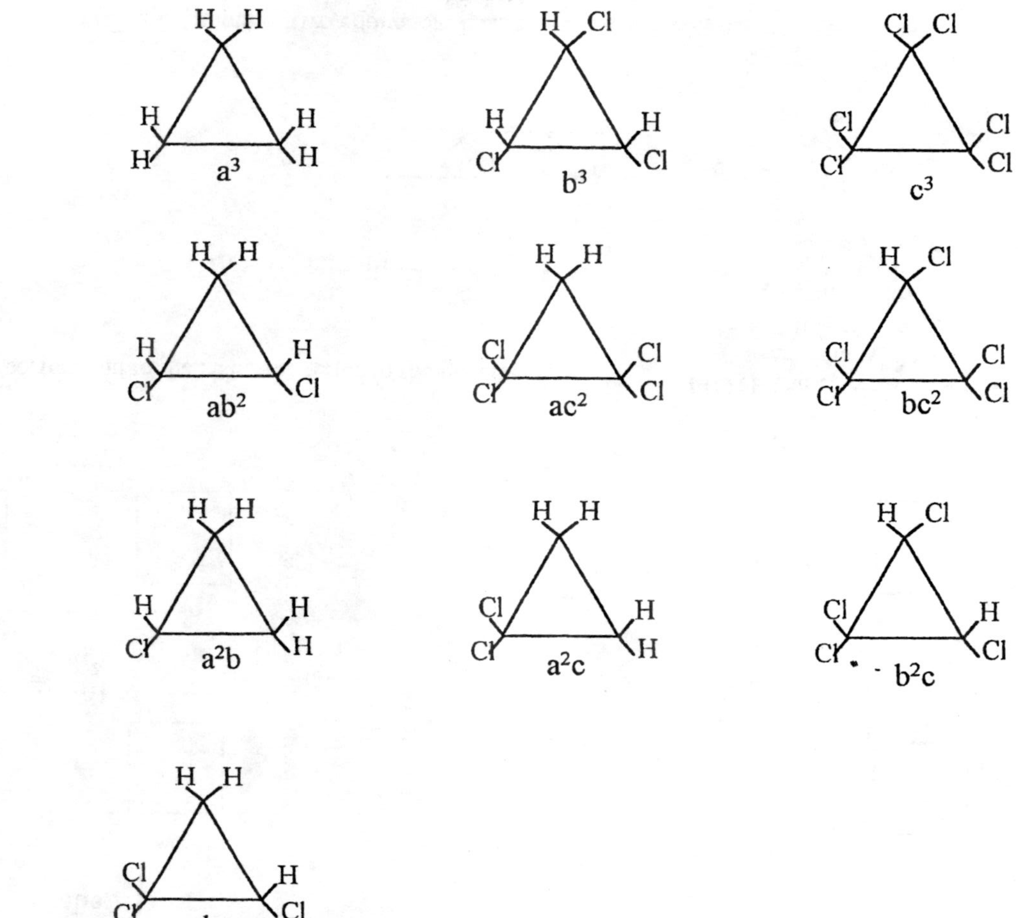

FIG9 The 10 distinct structures consistent with the general formula given by eqn.(21) and the generating function given by eqn.(25)

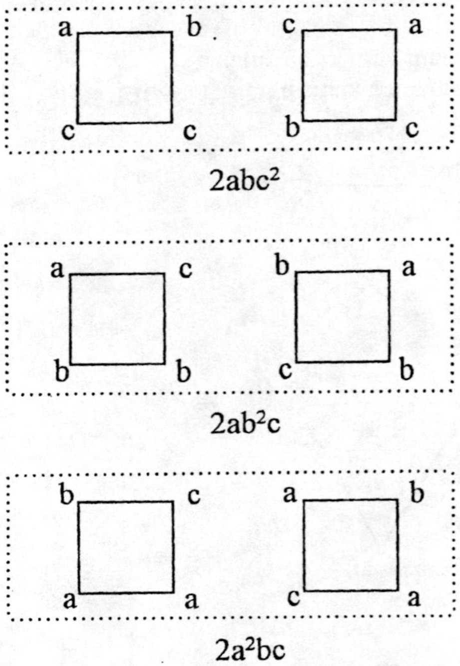

$2abc^2$

$2ab^2c$

$2a^2bc$

FIG10 The last six structures which correspond to the last three terms in the generating function given by eqn.(28)

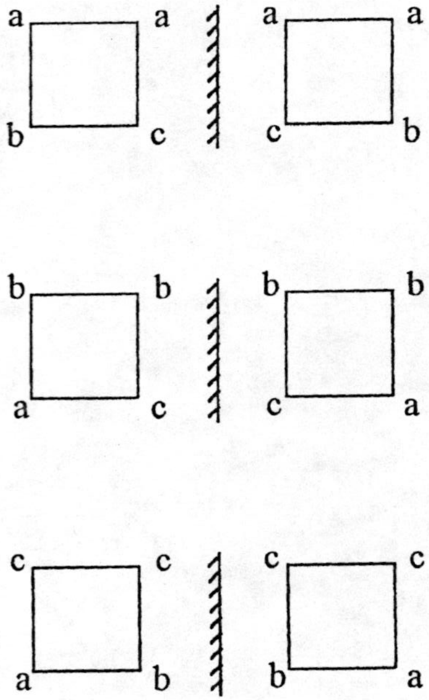

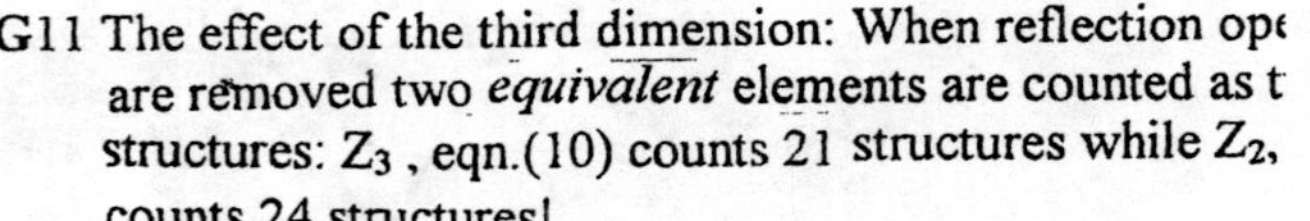

FIG11 The effect of the third dimension: When reflection opǝ are removed two *equivalent* elements are counted as t structures: Z_3 , eqn.(10) counts 21 structures while Z_2, counts 24 structures!

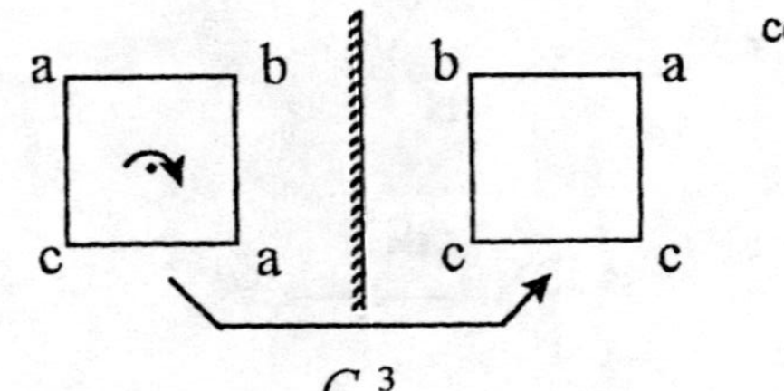

FIG12 Example of two equivalent elements counted by Z_2 as one structure because they are transformed to each other by C_4^3 which is present in Z_2.

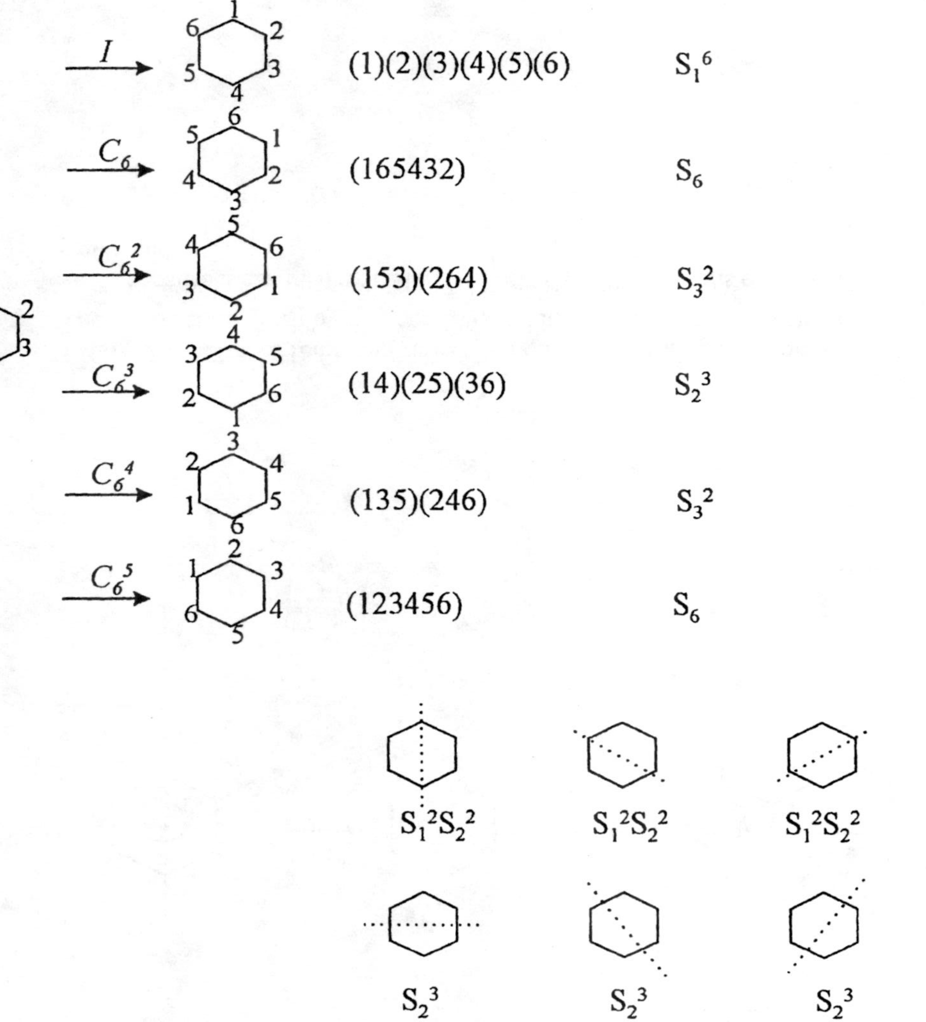

FIG13 The 12 symmetry operations (6 rotations and 6 reflections) of the regular hexagon: An example of the $\mathbf{D}_6$ point-group.

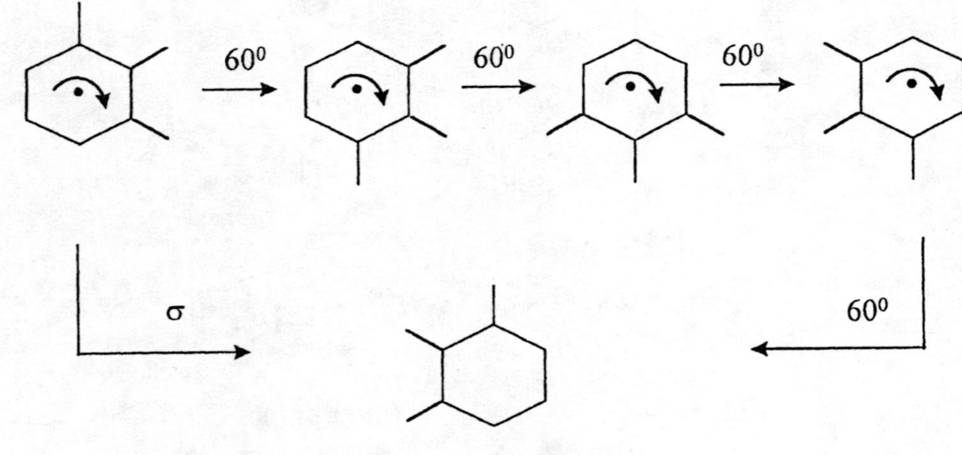

FIG14 The effect of removing reflections on the generating function [eqns.(37),(38)]. The structure given by ii in eqn.(39) and its mirror-image are counted as one structure in Z_2 because it can be obtained by 4 successive 60° rotations on ii which is contained in Z_2.

A. Rotations

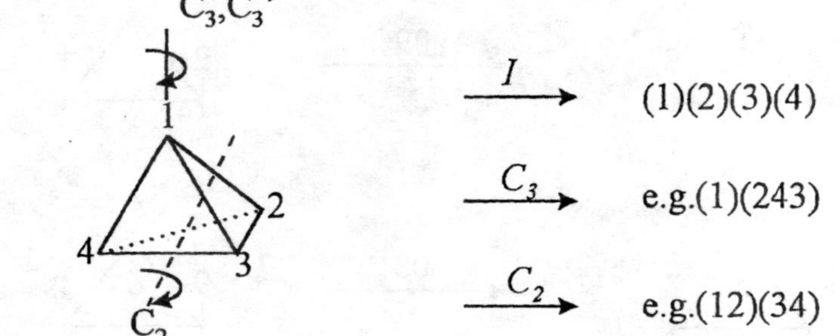

$\xrightarrow{I}$	(1)(2)(3)(4)	S_1^{4}
$\xrightarrow{C_3}$	e.g.(1)(243)	$8S_1^{1}S_3^{1}$
$\xrightarrow{C_2}$	e.g.(12)(34)	$3S_2^{2}$

B. Reflecions

S_4 operations (illustrated by inscribing the tetrahedron in a cube)

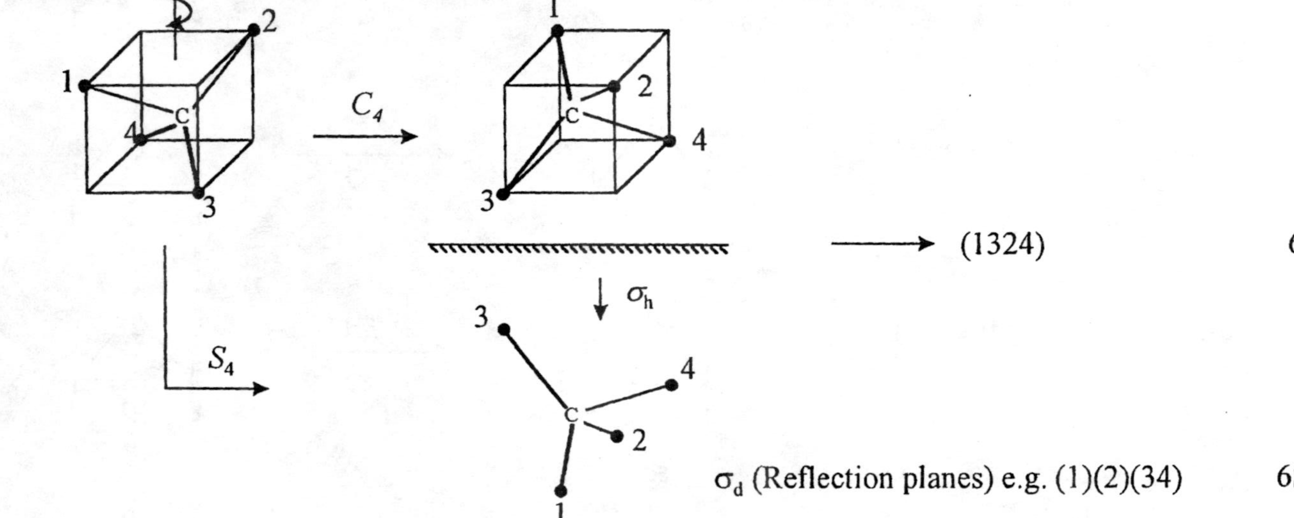

$\longrightarrow$ (1324) $6S_4^{1}$

σ_d (Reflection planes) e.g. (1)(2)(34) $6S_1^{2}S_2^{1}$

FIG15 The symmetry operations of the tetrahedron , an example of the Td point-group.

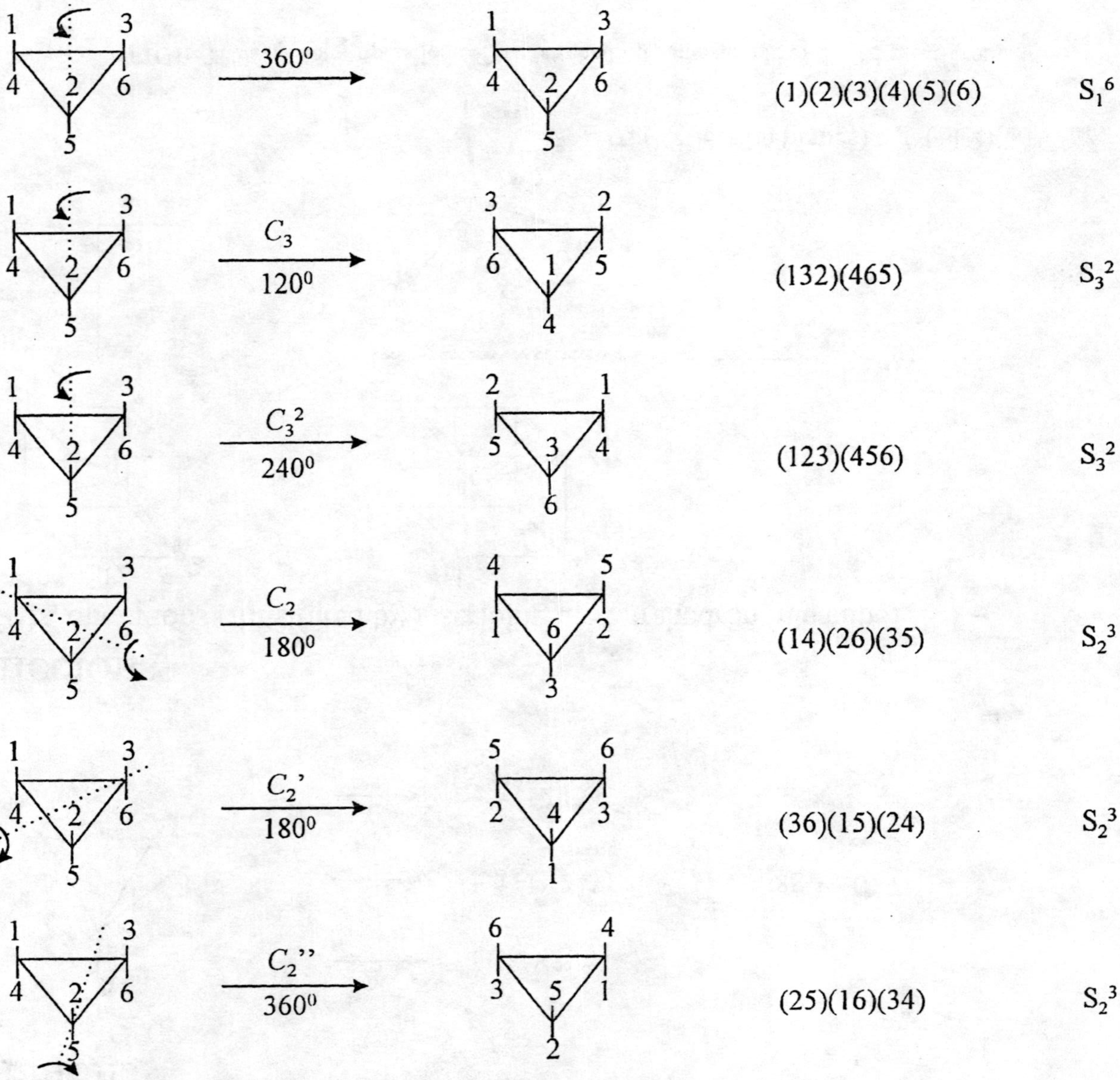

FIG16 Rotation symmetry operations of Cyclopropane , c.f.the $\mathbf{D_3}$ point group.

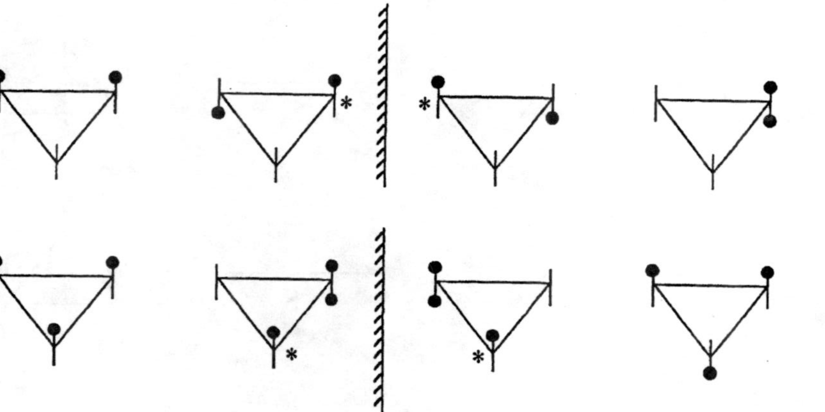

FIG17 Di- and tri- substituted cyclopropane derivatives, eqn.(51) generated using Z_2. The two middle structures are enantiomers . Chiral atoms are marked with asterices. When Z_3 is used the two enantiomers are counted as one (distinct) structure.

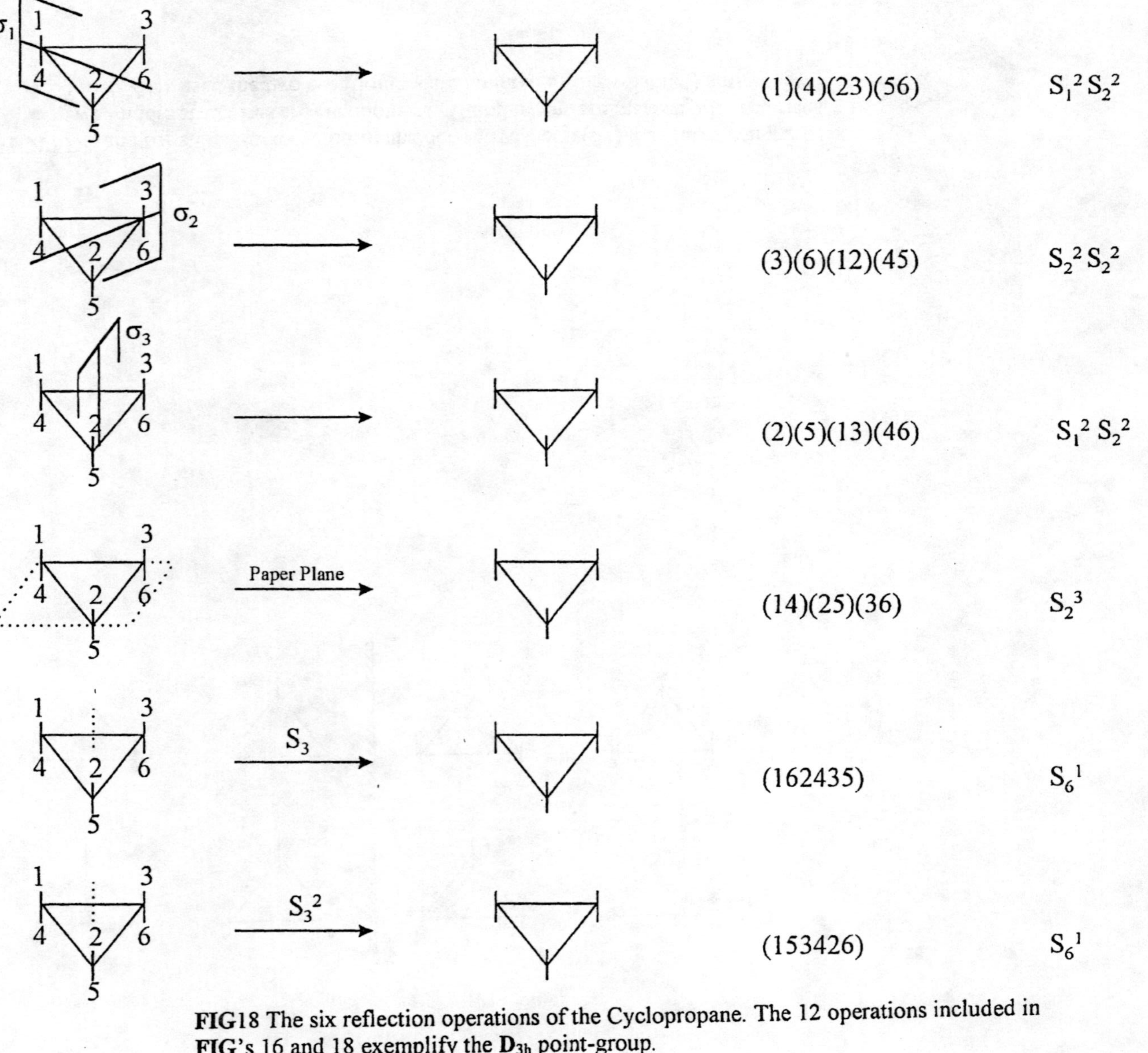

FIG18 The six reflection operations of the Cyclopropane. The 12 operations included in **FIG**'s 16 and 18 exemplify the $\mathbf{D_{3h}}$ point-group.

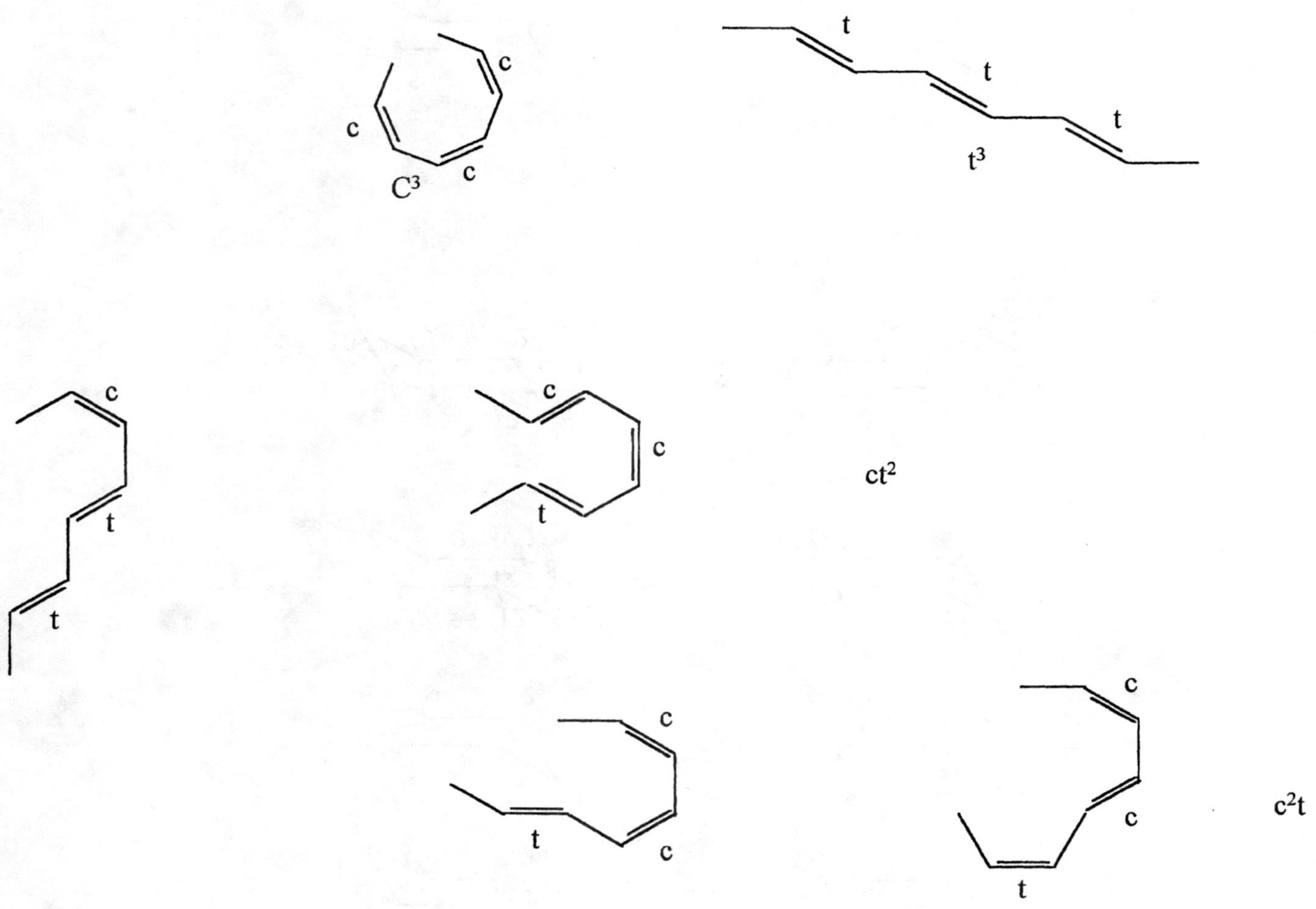

C^3

t^3

ct^2

c^2t

FIG19 The six cis-trans-isomers of 1,6-dimethyl 1,3,5-hexatriene.

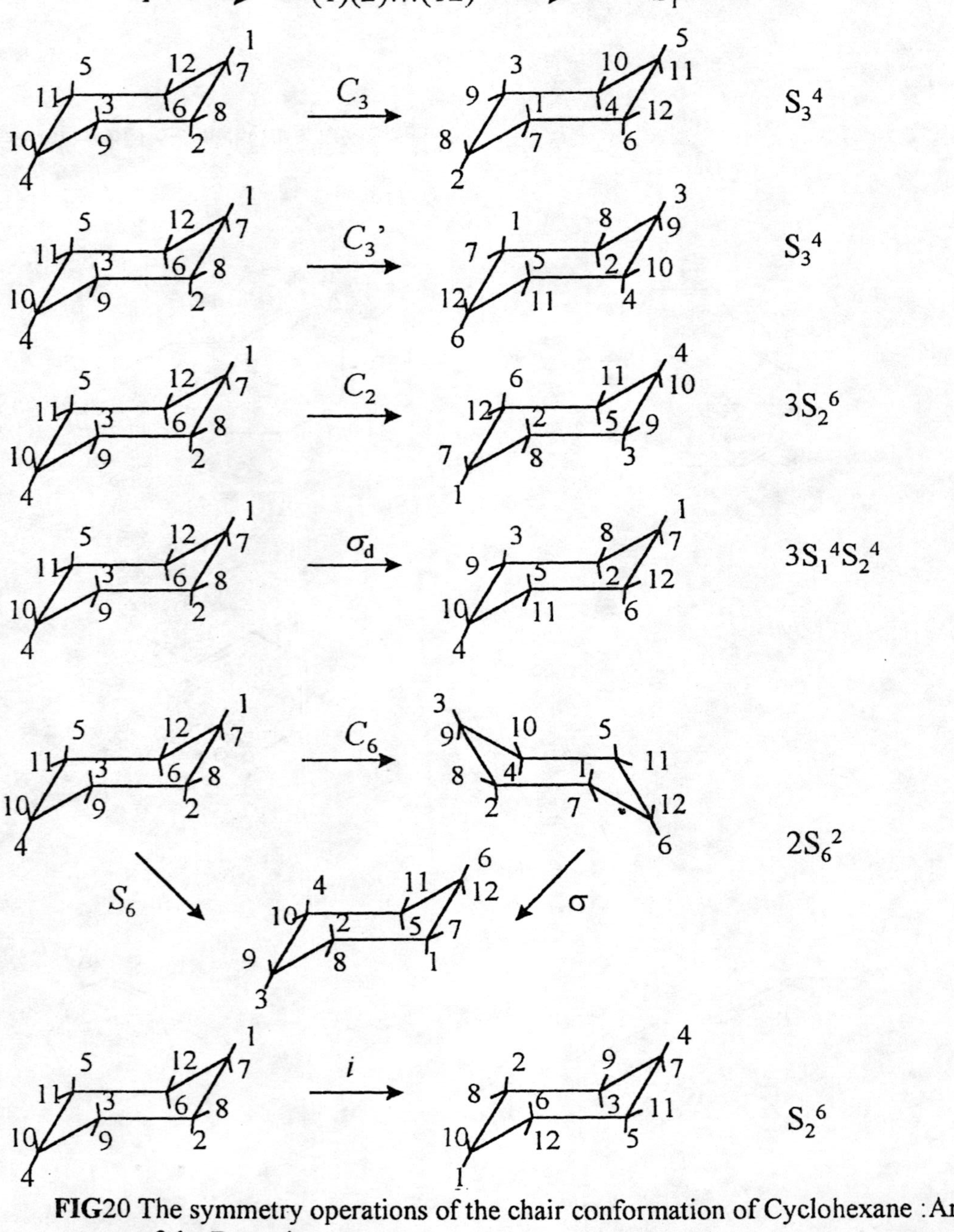

FIG20 The symmetry operations of the chair conformation of Cyclohexane : An example of the $\mathbf{D_{3d}}$ point-group.

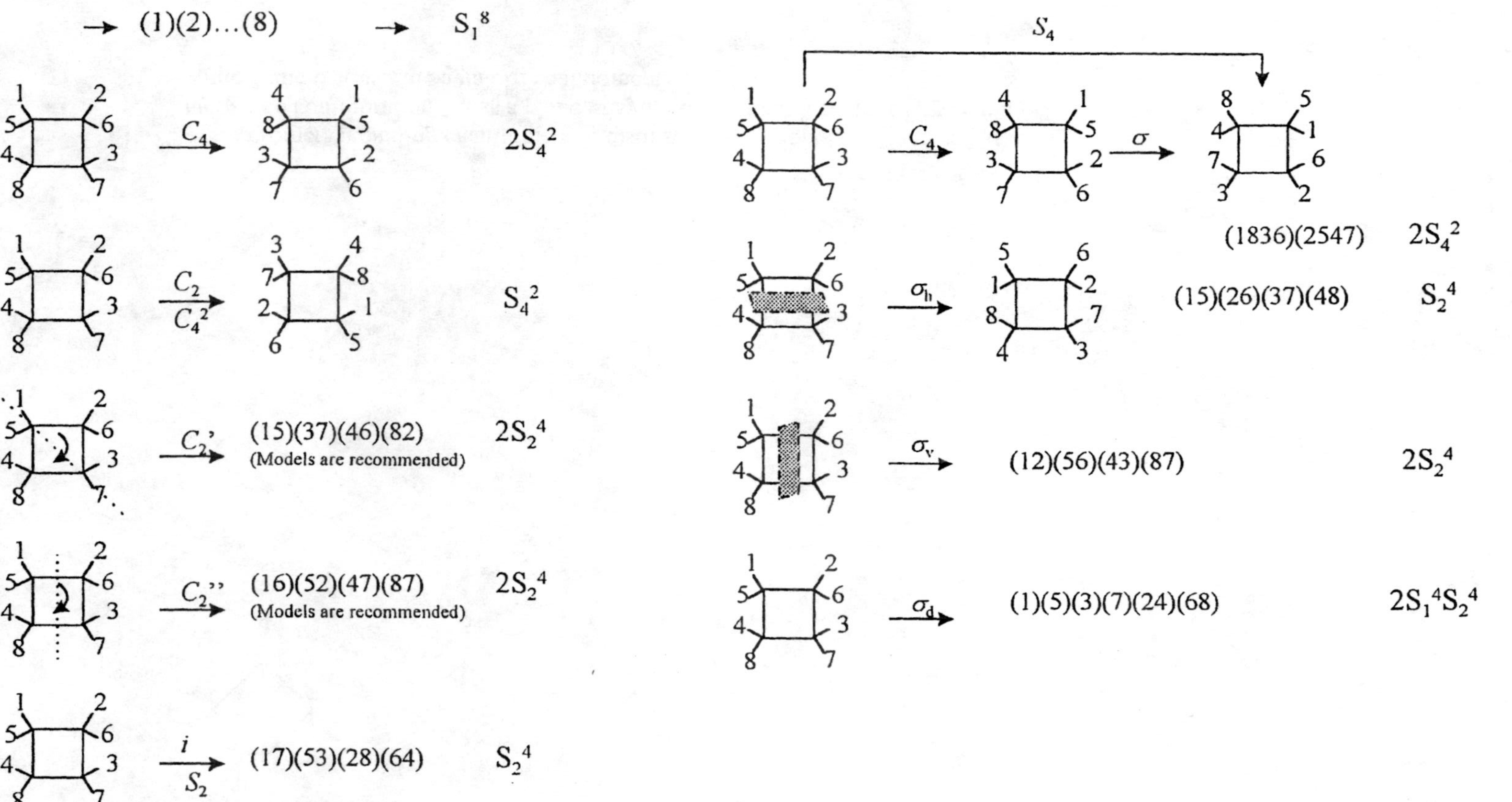

FIG21 The symmetry operations of unsubstituted cyclobutane , $\mathbf{D_{4h}}$ point-group.

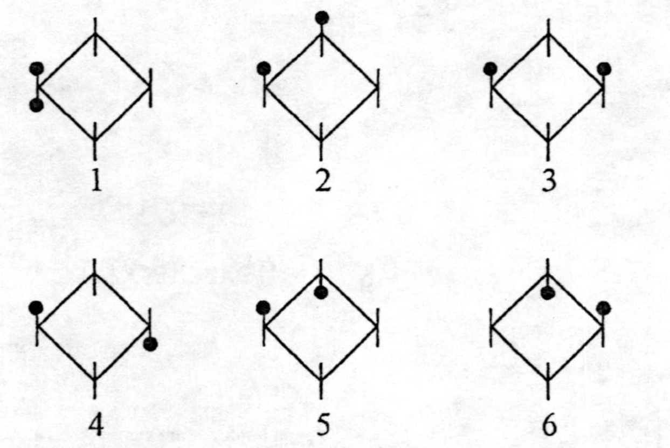

FIG22 Stereoisomers including enantiomers of disubstituted cyclobutanes.Isomer 2 is a *meso*-like configuration, 3 has a plane of symmetry , 4 has a center of inversion while 5 and 6 constitute a pair of enantiomers.

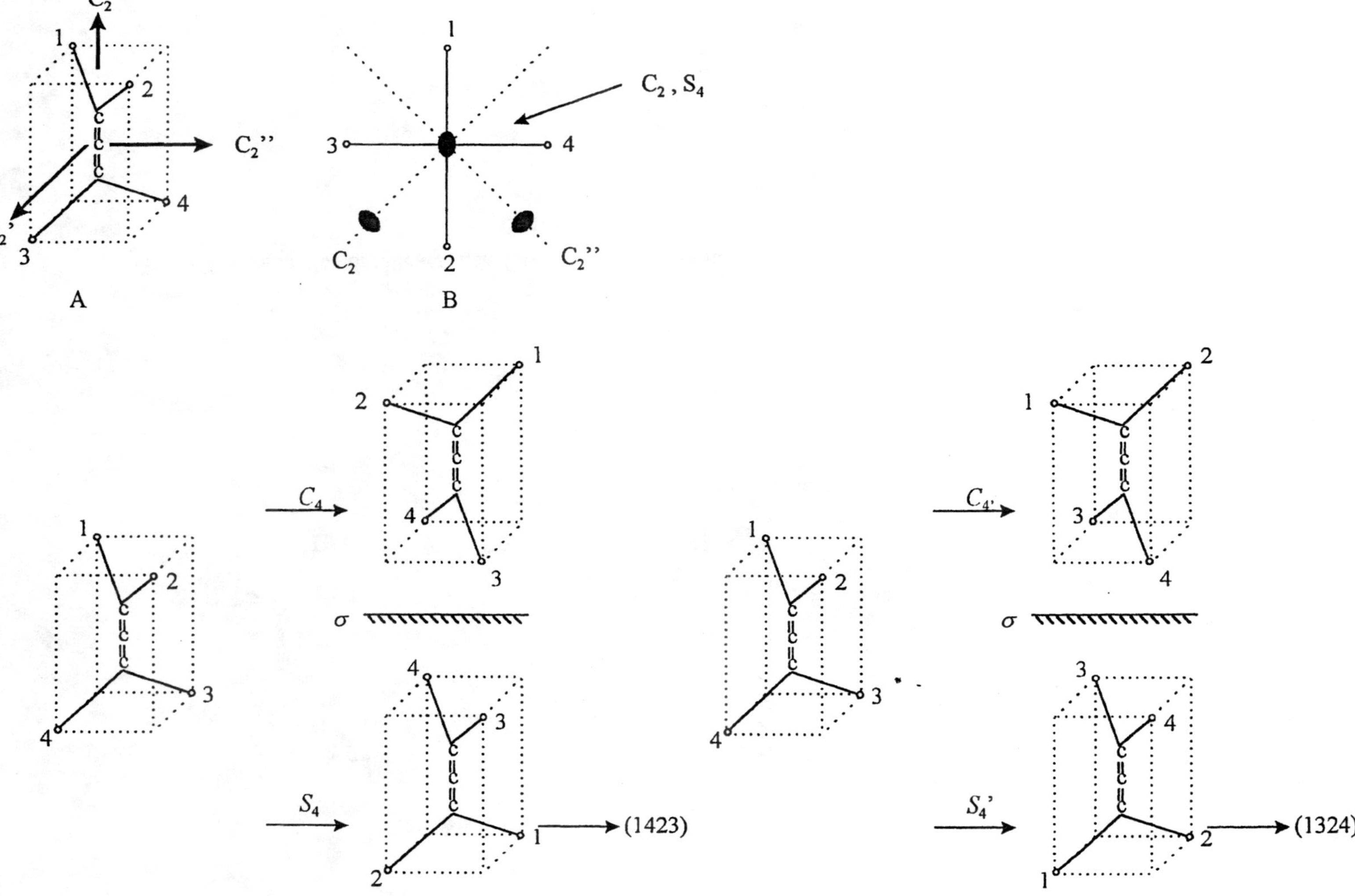

FIG23 Allene inscribed in an elongated cube, A and a top view ,B. The S_4 and S_4' operations are also demonstrated.

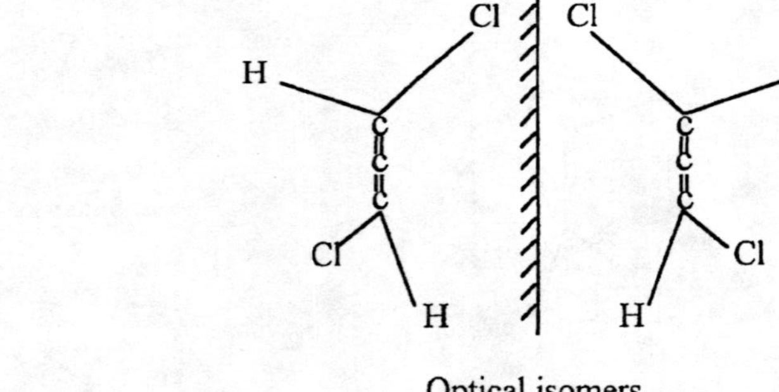

FIG24 The three dichloroallenes including a pair of enantiomers.

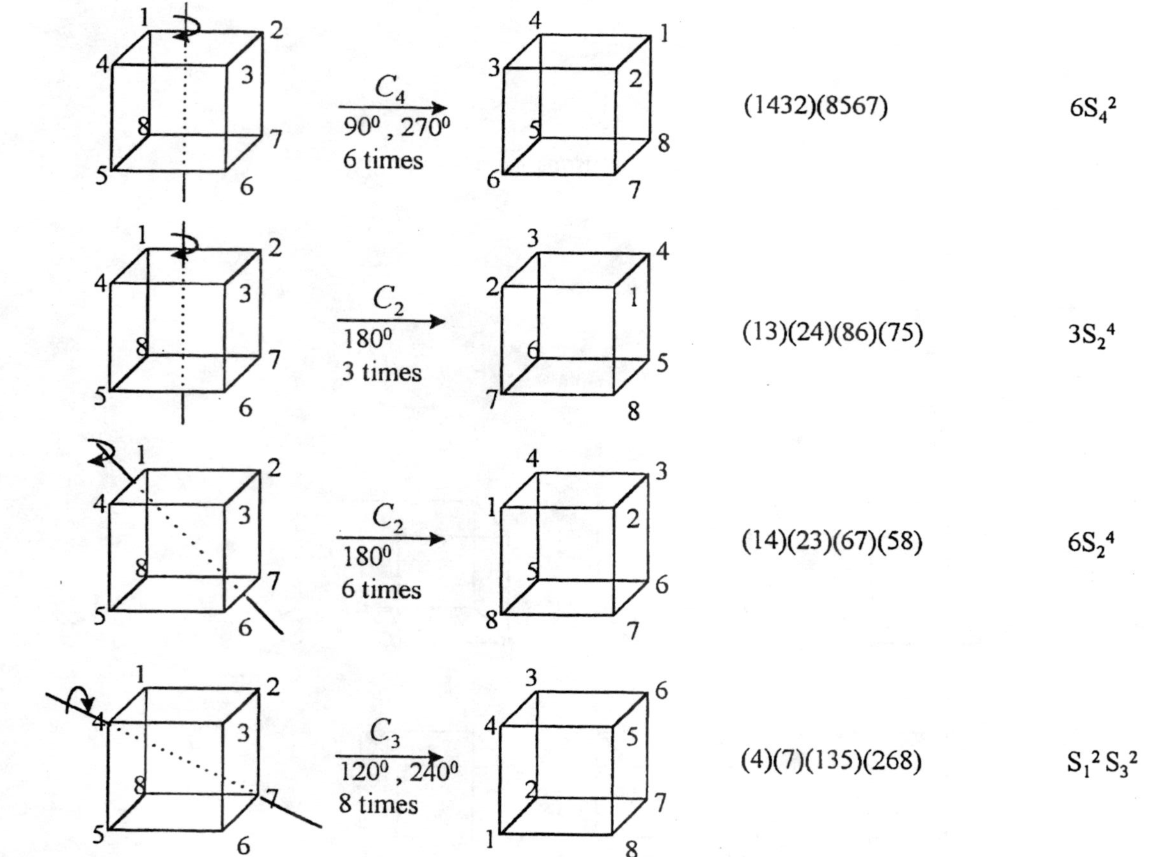

FIG25 Symmetry operations and their permutation representations of the cube, the **O** point-group.

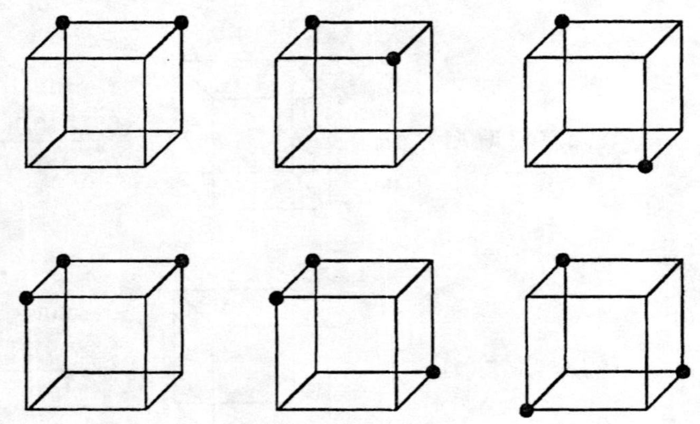

FIG26 Di- and tri- substituted cubanes.

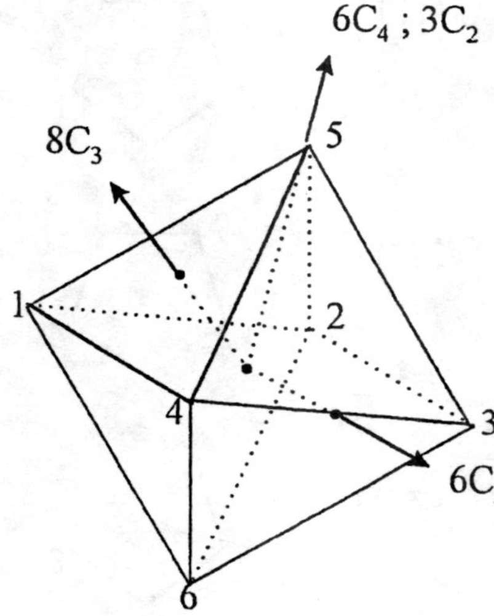

FIG27 Symmetry elements of the octahedron.

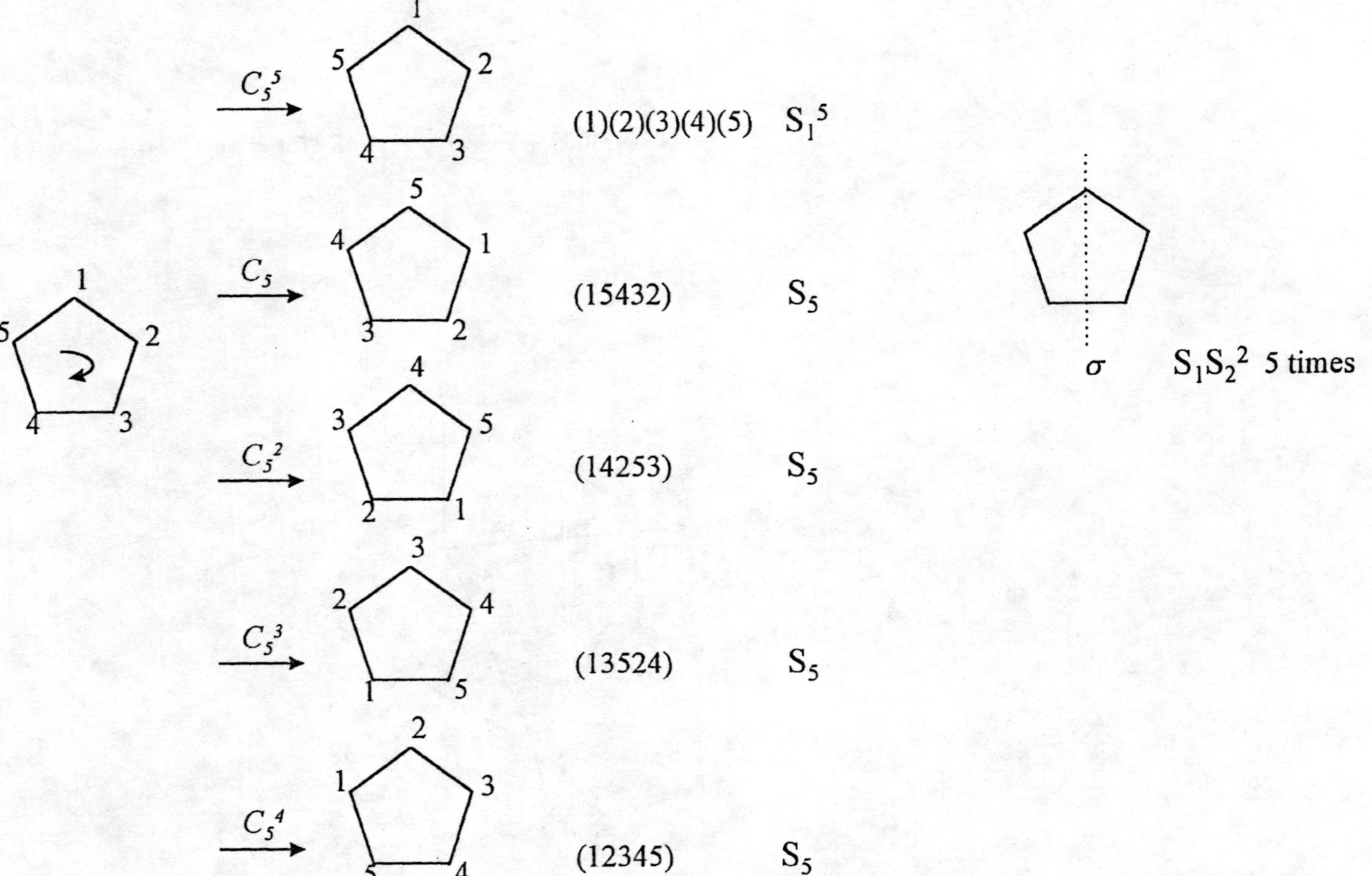

FIG28 Symmetry operations of the regular pentagon.

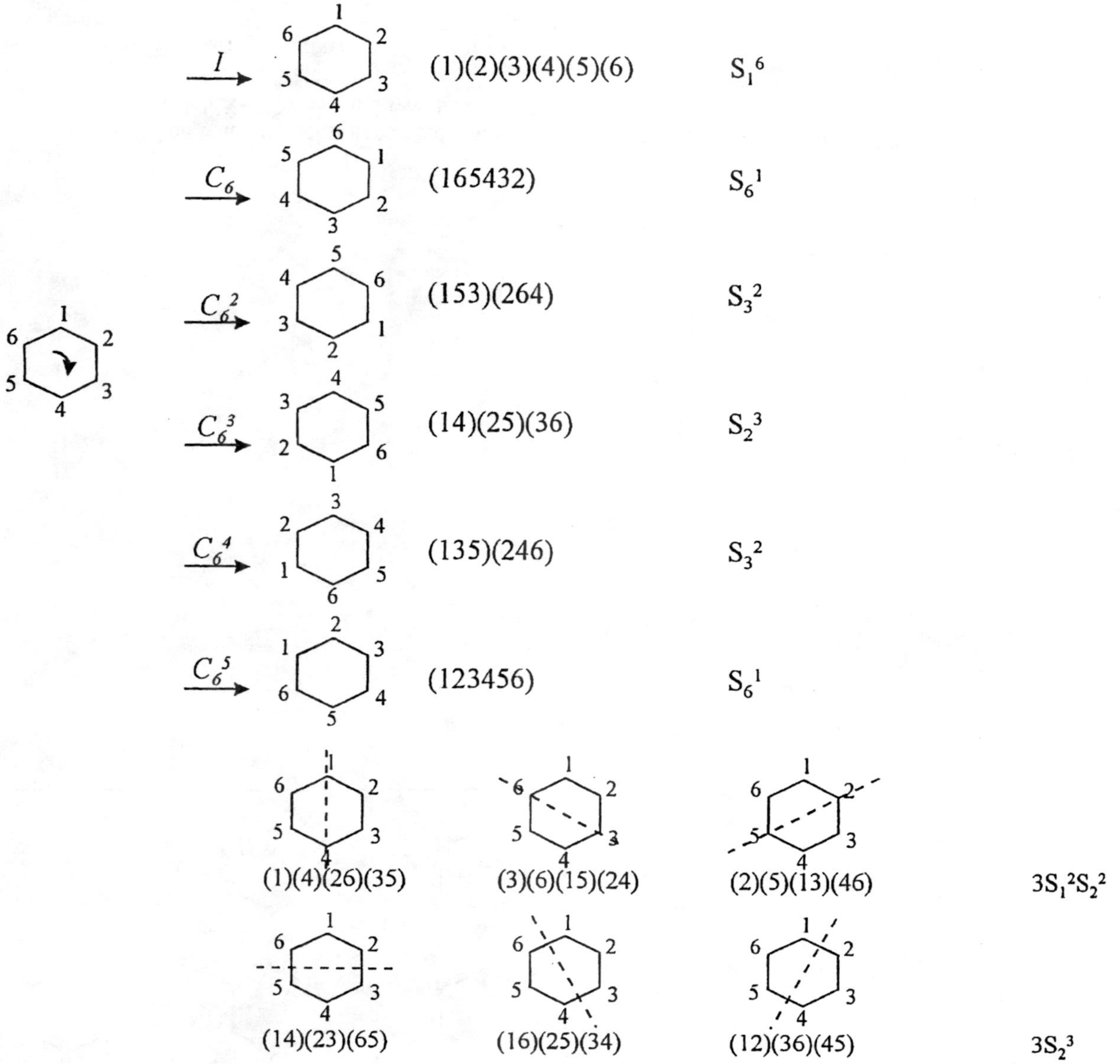

Operation	Permutation	Cycle index term
I	(1)(2)(3)(4)(5)(6)	S_1^6
C_6	(165432)	S_6^1
C_6^2	(153)(264)	S_3^2
C_6^3	(14)(25)(36)	S_2^3
C_6^4	(135)(246)	S_3^2
C_6^5	(123456)	S_6^1

(1)(4)(26)(35) (3)(6)(15)(24) (2)(5)(13)(46) $3S_1^2S_2^2$

(14)(23)(65) (16)(25)(34) (12)(36)(45) $3S_2^3$

FIG29 Symmetry operations of the regular hexagon , $\mathbf{D_6}$ point-group.

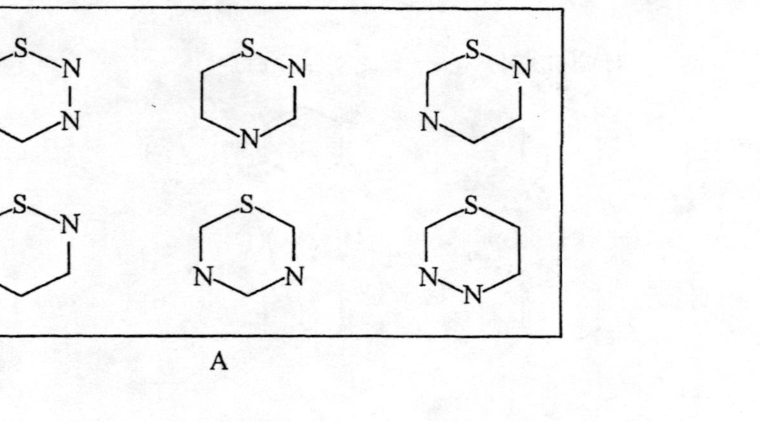

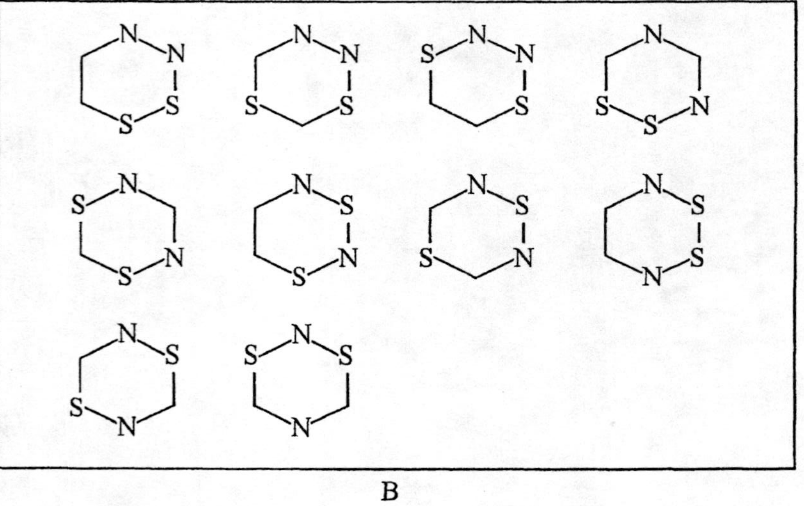

FIG30 The 6 heterocycles generated from the general system shown in eqn.(90) for which X={C,N,S} and which contain
A)2N and 1S atoms
B)2N and 2S atoms

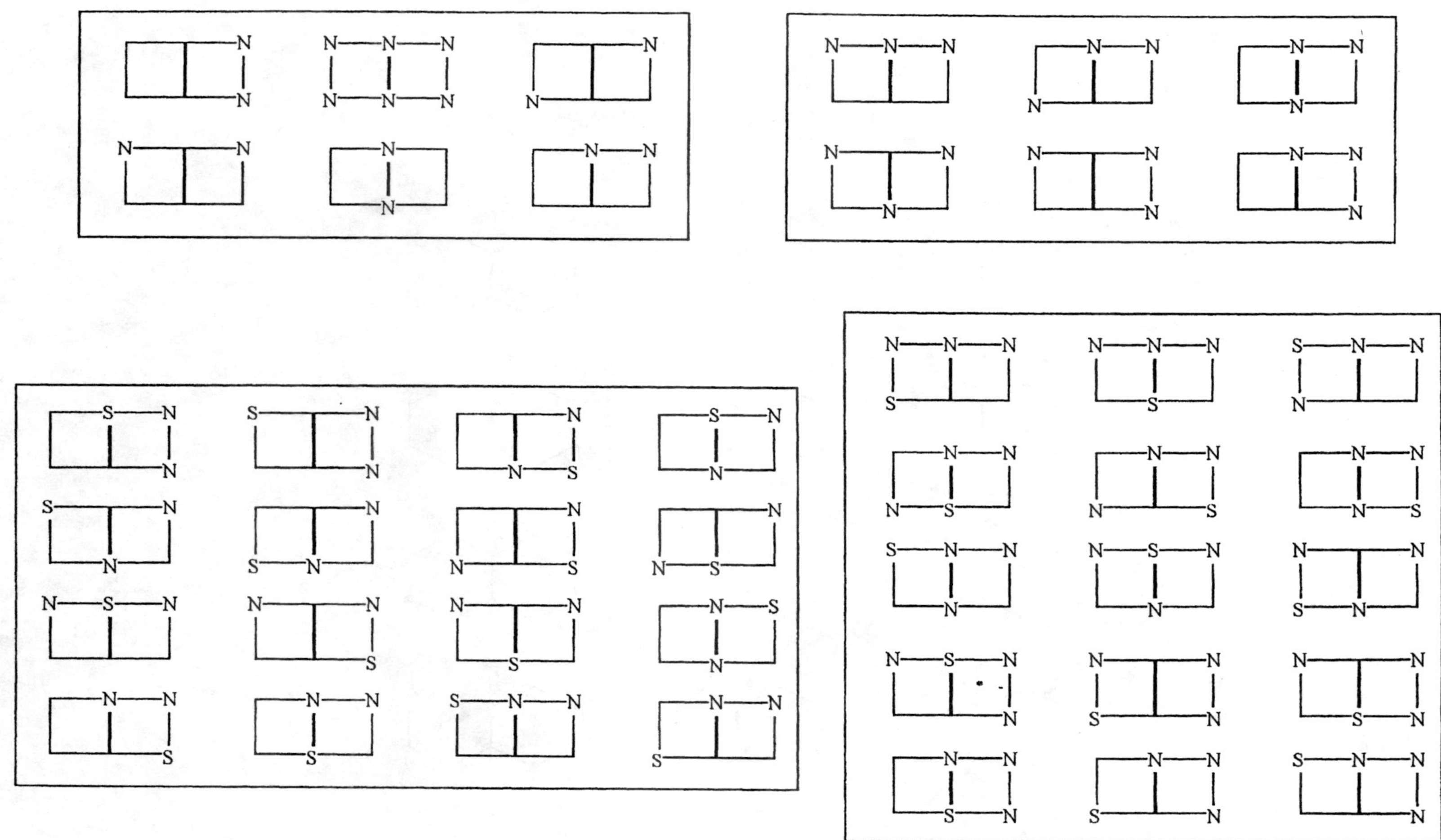

FIG31 Heterocyclic compounds generated from the molecular graph of bicyclobutadiene [eqns.(96),(97)].

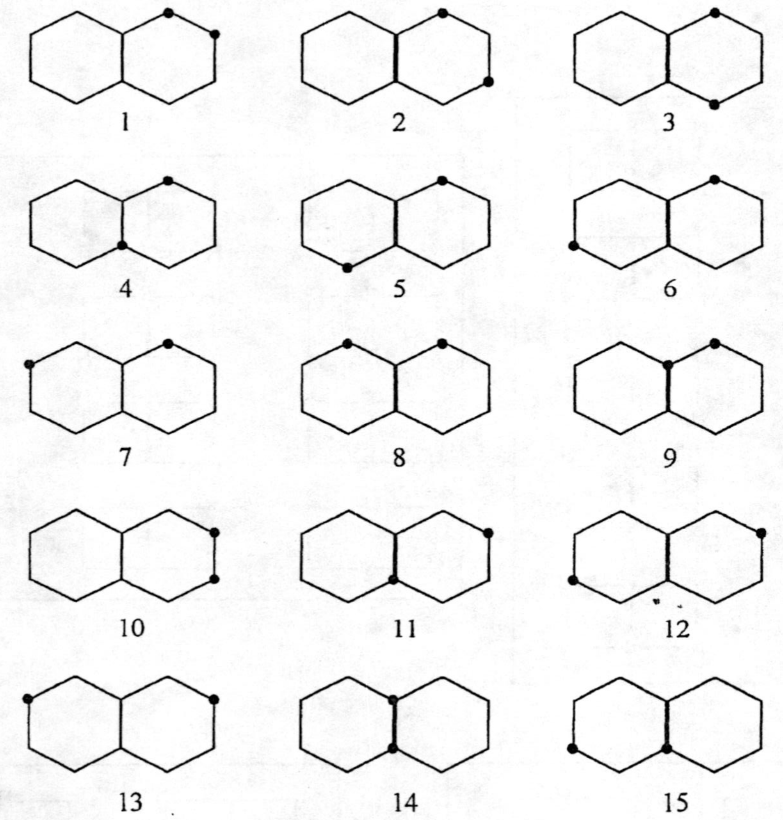

FIG32 Heterocyclic systems generated from system3,Table 2

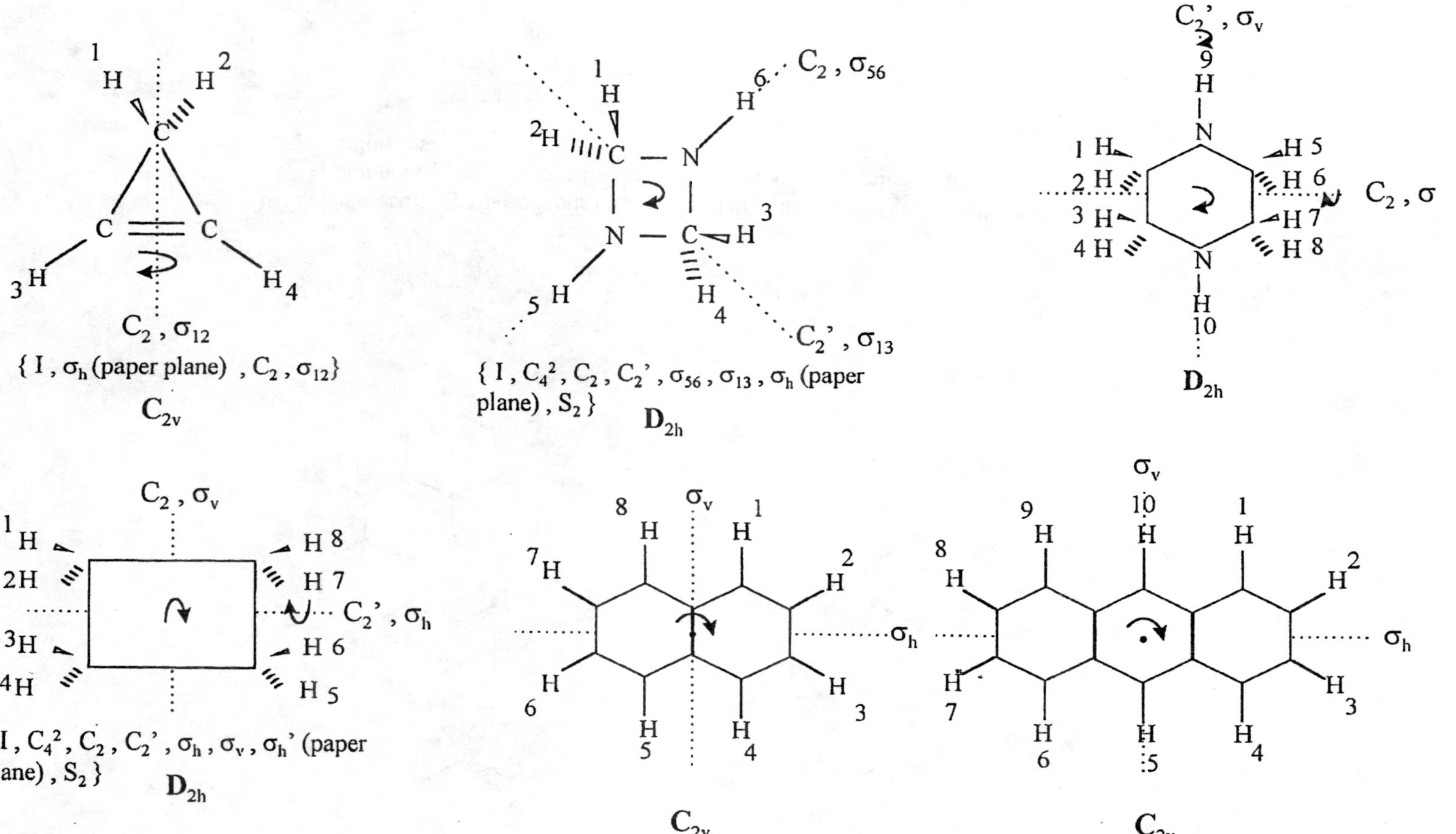

FIG33 Six chemical systems designed to illustrate how to use the formula of Balasubramanian to find pmr [eqn.(98)]

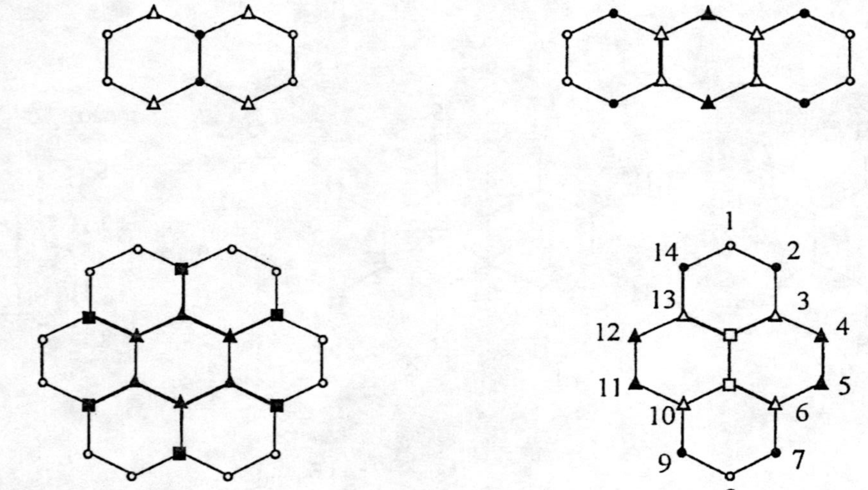

FIG34 Molecular graphs of four benzenoid hydrocarbons designed to illustrate computation of ^{13}C nmr.Orbits of C atoms are distinguished by various types of colorings.

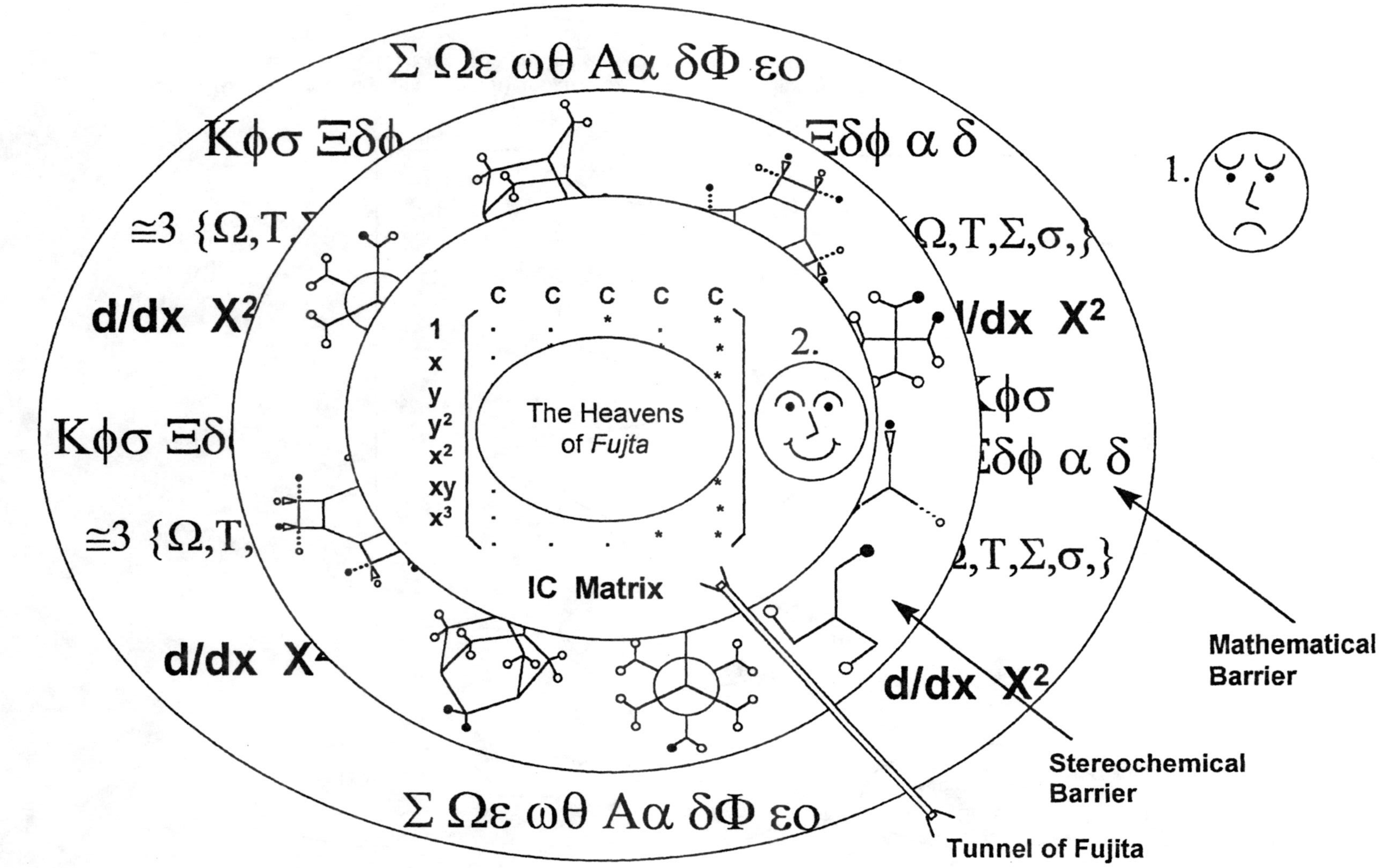

1. An organic chemist trying to make his way through !
2. He made it.

FIG35 Pictorial illustration of a paradox: An organic chemist (who is not interested in mathematics) is trying to make his way through the "heavens of Fujita". Such heavens can only be reached by mathematics! C.f. comments a-c in Section 3.1

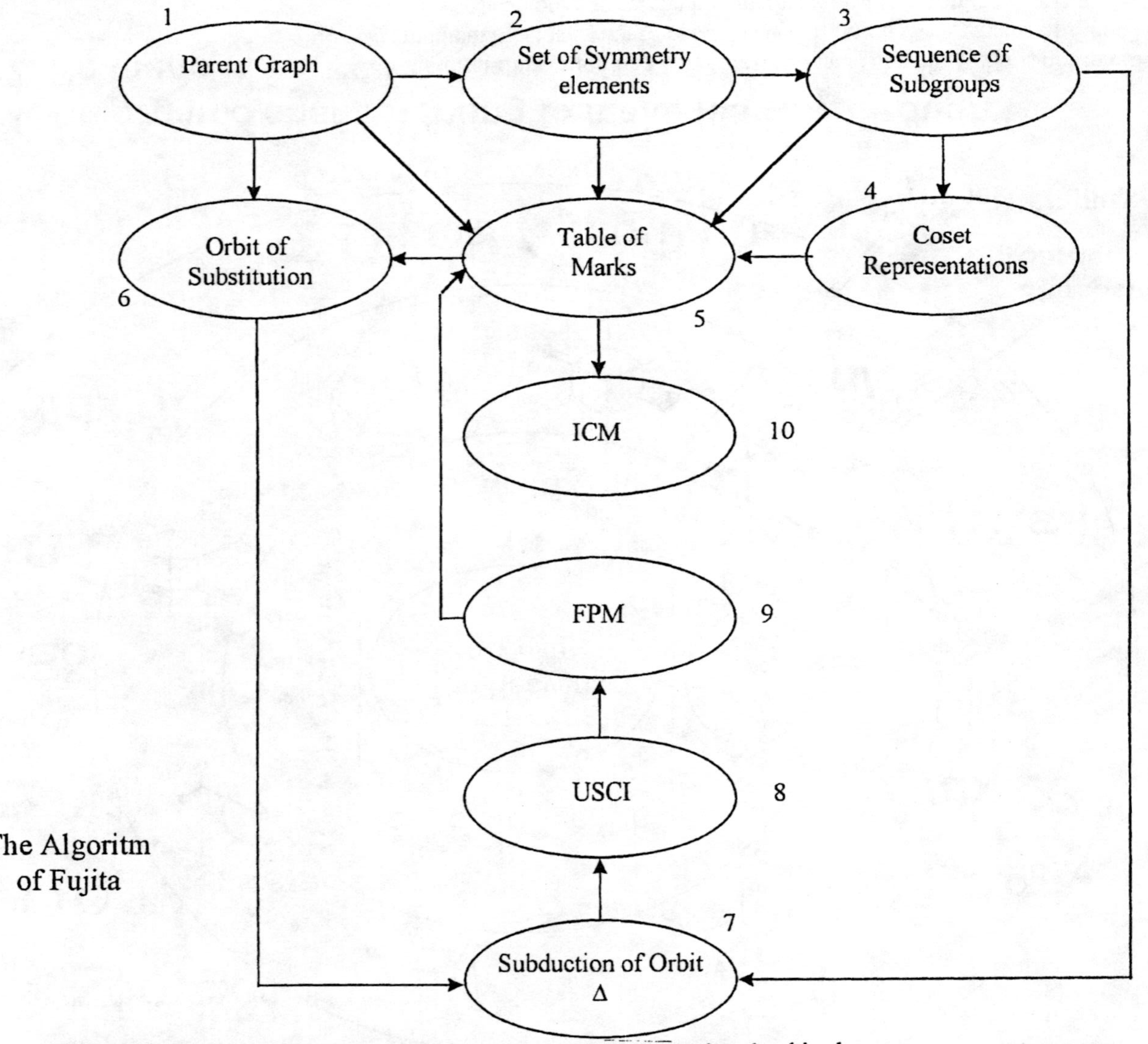

FIG36 A directed graph (digraph) designed to define the steps involved in the enumeration scheme of Fujita and show how these steps are inter-related to each other.

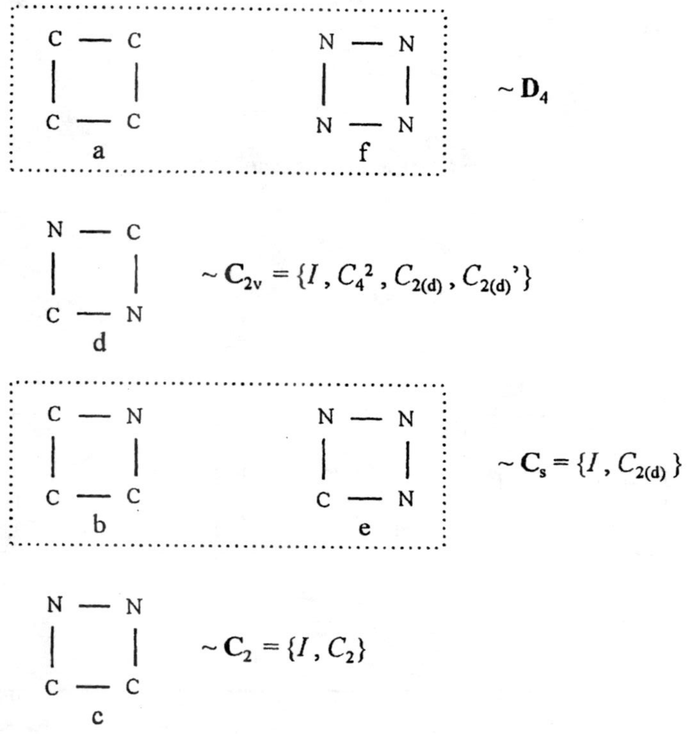

```
C — C          N — N
|   |          |   |      ~ D4
C — C          N — N
  a              f

N — C
|   |      ~ C2v = {I, C4^2, C2(d), C2(d)'}
C — N
  d

C — N          N — N
|   |          |   |      ~ Cs = {I, C2(d)}
C — C          C — N
  b              e

N — N
|   |      ~ C2 = {I, C2}
C — C
  c
```

$\sim \mathbf{D}_4$

$\sim \mathbf{C}_{2v} = \{I, C_4^2, C_{2(d)}, C_{2(d)}'\}$

$\sim \mathbf{C}_s = \{I, C_{2(d)}\}$

$\sim \mathbf{C}_2 = \{I, C_2\}$

FIG37 Modeling some of the subgroups of the $\mathbf{D}_4$ point-group. Each structure remains fixed only under the operations of the indicated subgroup elements.

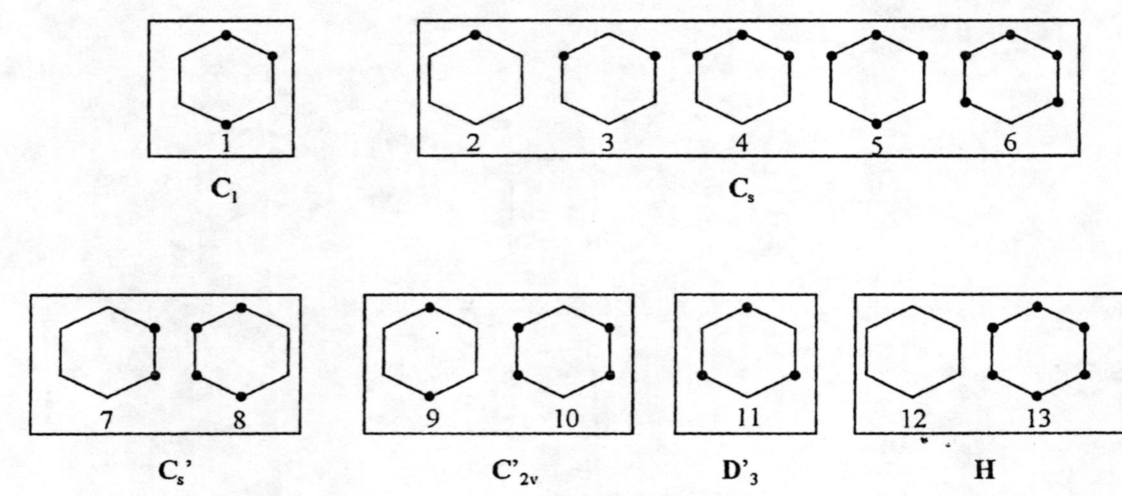

FIG38 Heterocycles which correspond to the isomer-count matrix, Table 23.

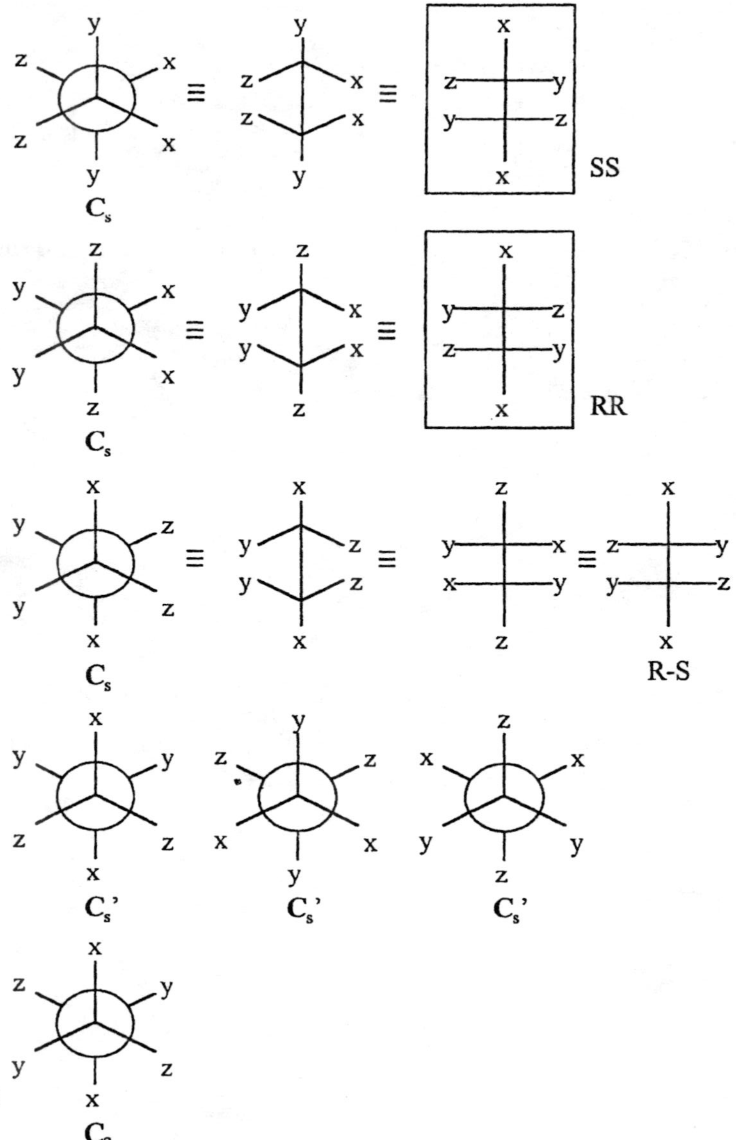

FIG39 Isomers which correspond to $x^2y^2z^2$ in the isomer-count matrix , Table26.RR,SS and RS (meso-) tartaric acids are identified .

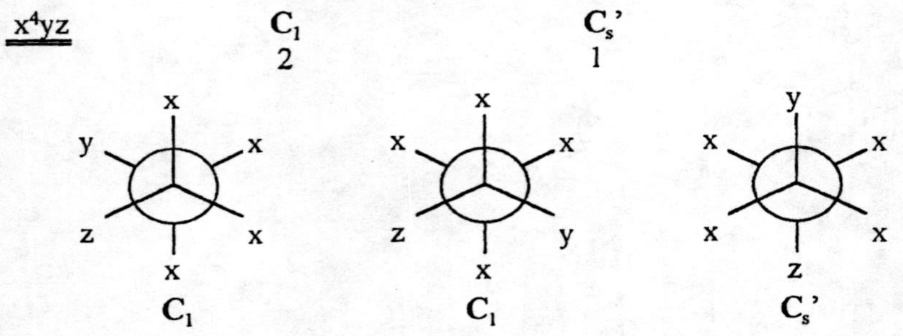

$\underline{x^4yz}$ C_1 2 C_s' 1

x, y, x, z, x, x — C_1

x, x, x, z, x, y — C_1

y, x, x, x, z, x — C_s'

FIG40 Isomers which correspond to x^4yz in the isomer-count matrix . Table 26.

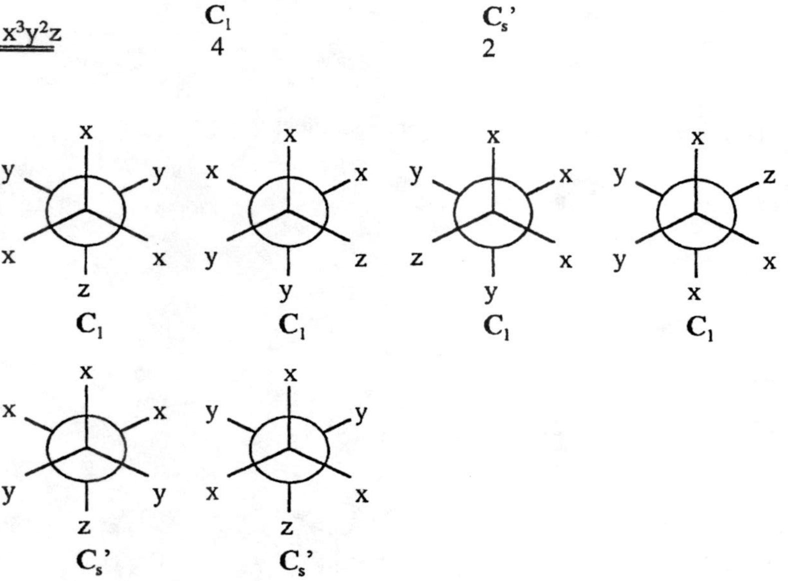

FIG41 Isomers which correspond to x^3y^2z in the ICM , Table 26.

x^3y^3	C_1	C_s'	C_{3v}
	1	1	1

C_1 C_s' C_{3v}

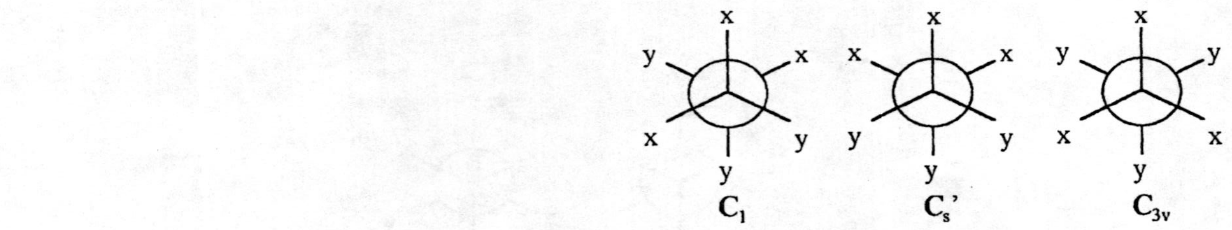

FIG42 Isomers which correspond to x^3y^3 in the ICM , Table 26.

$\underline{x^4y^2}$

C_s	C_s'	C_{2v}
1	1	1

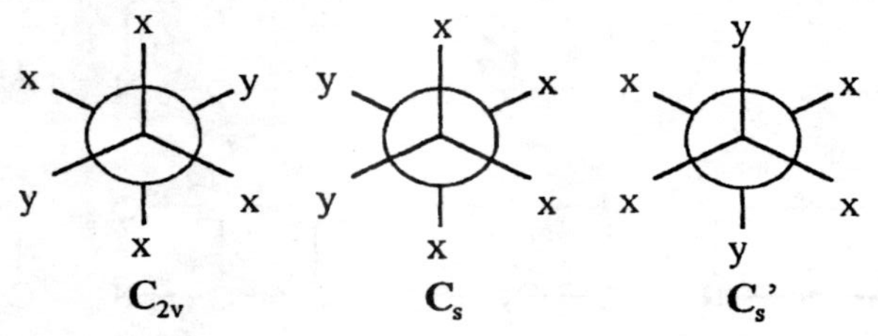

FIG43 Isomers which correspond to x^4y^2 in the ICM , Table 26.

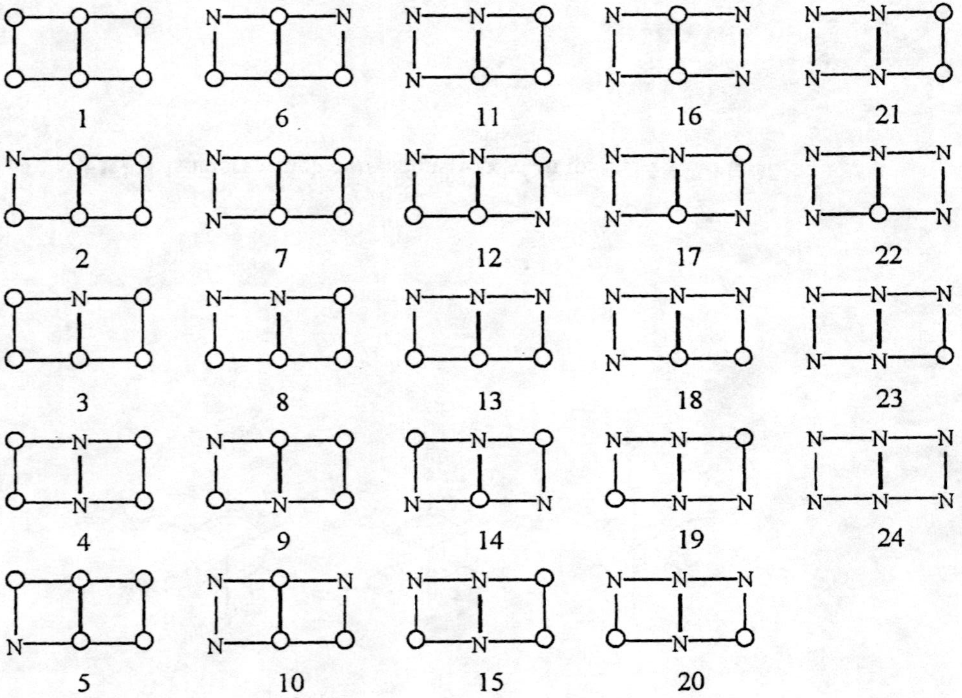

FIG44 Heterocyclic compounds labeled in the isomer-count matrix , Table 29.

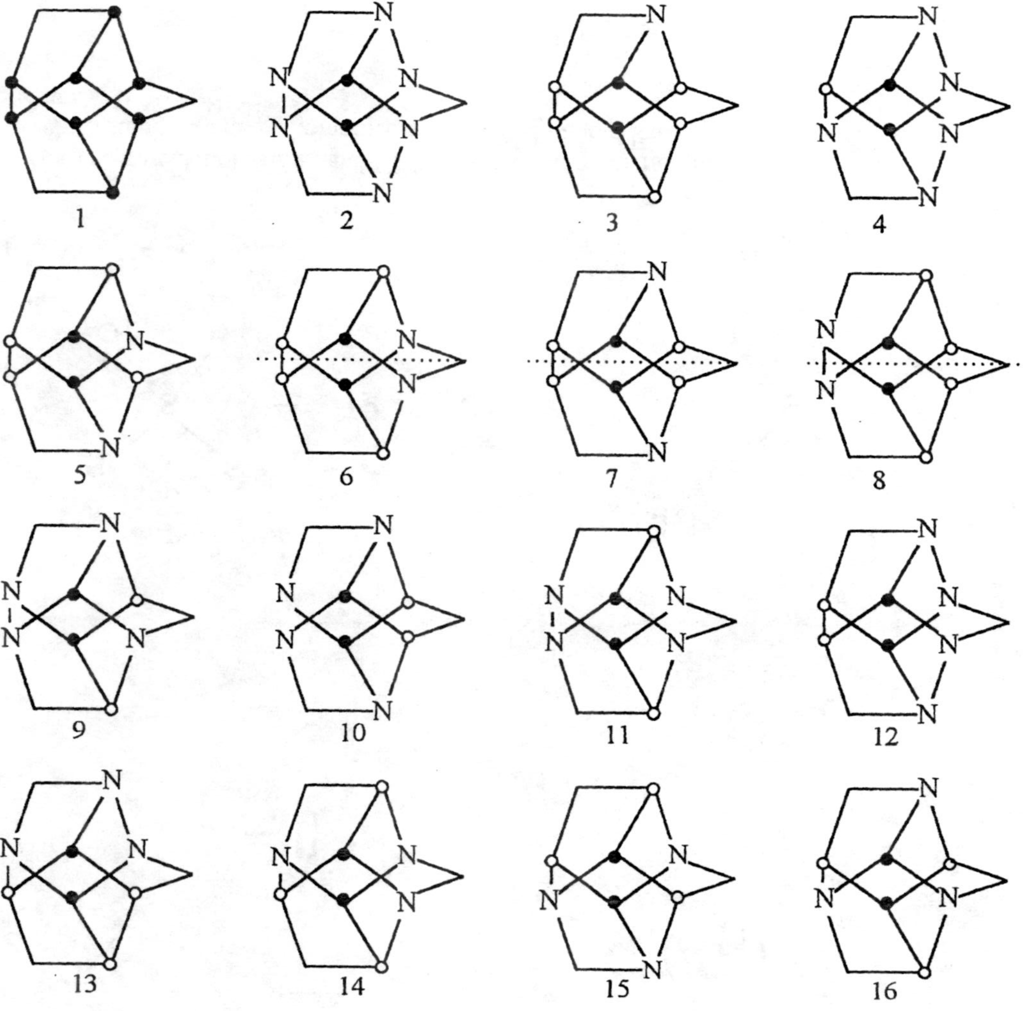

FIG45 Heterocyclic compounds derived from trishomocubane which correspond to the ICM . Table 32.

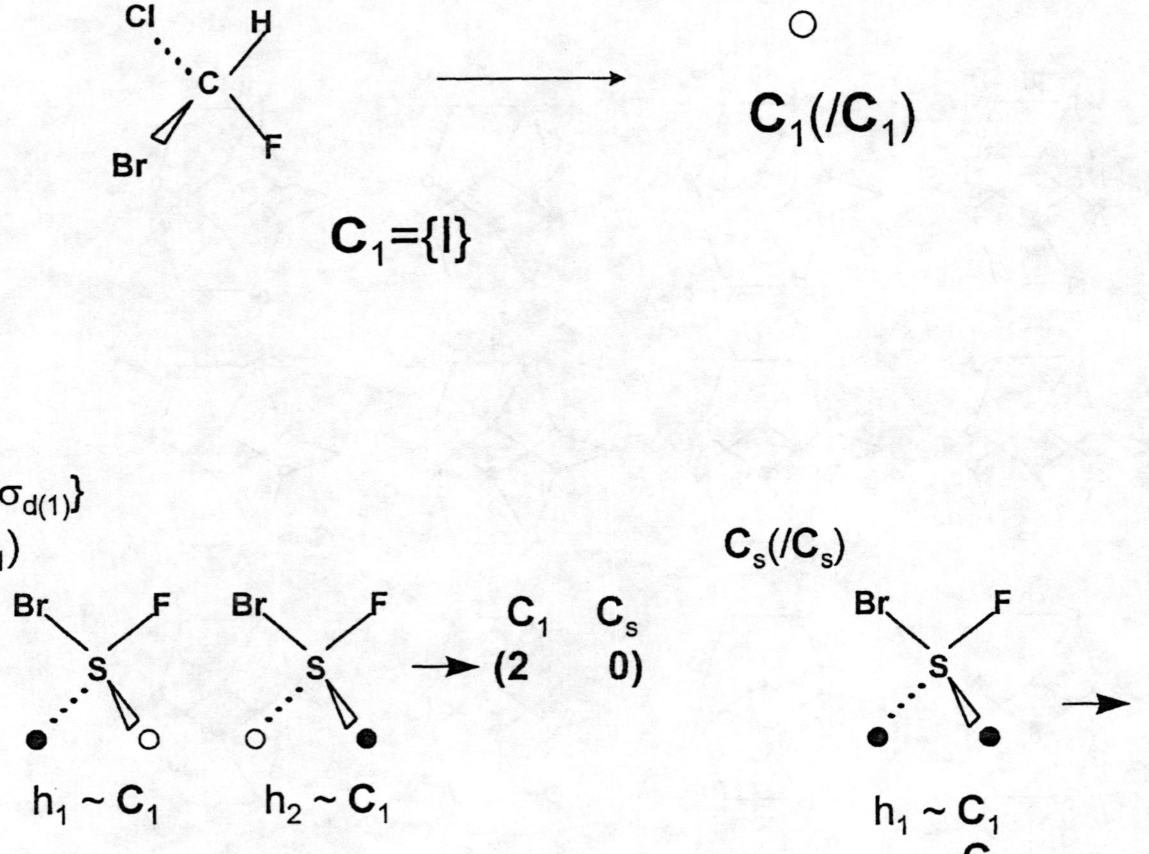

FIG46 Modeling coset representations of $\mathbf{C_1}$ and $\mathbf{C_s}$. The h_i's are homomers. The subgroups which stabilize a given homomer are indicated in bold letters. Mark rows are given.

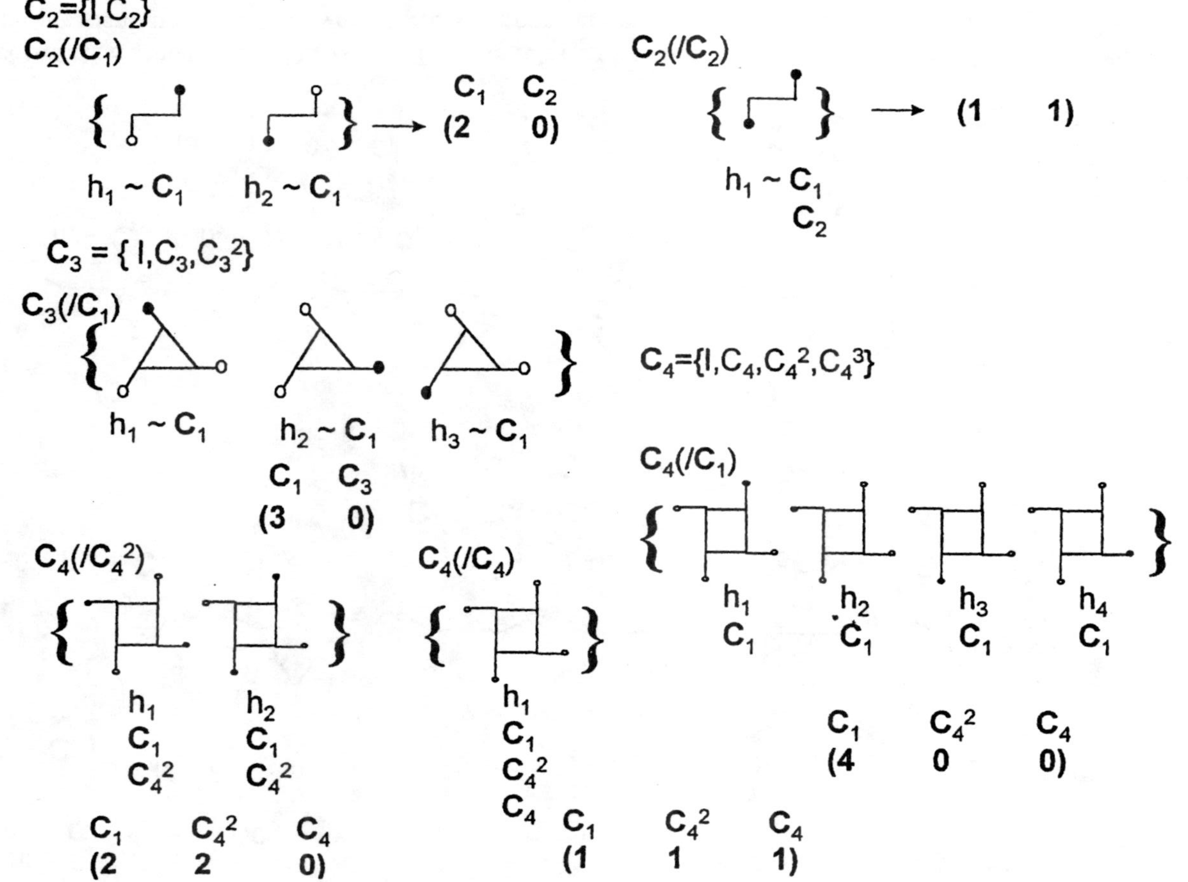

FIG47 Homomers which model corset representations of C_2, C_3 and C_4 and subgroups stabilize them. Mark rows are found by listing the numbers of such groups.

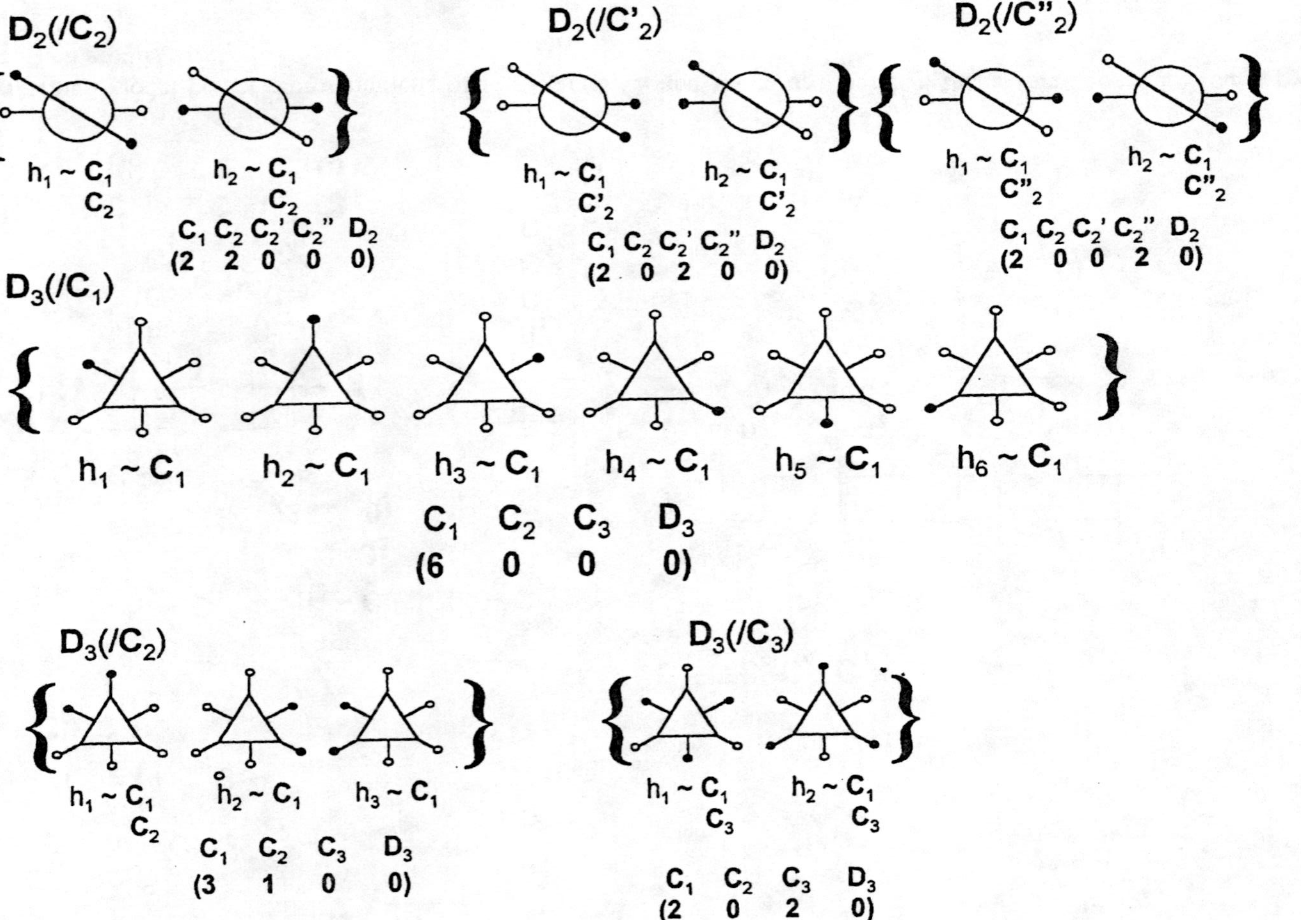

FIG48 Modelling corset representations of $\mathbf{D_2}$, $\mathbf{D_3}$ and $\mathbf{D_4}$. The relevant homomers are drawn as colored graphs. Subgroups which leave each colored graph invariant are indicated. When these subgroups are counted and the resulting counts are listed in the order of subgroup-sequence mark rows result.

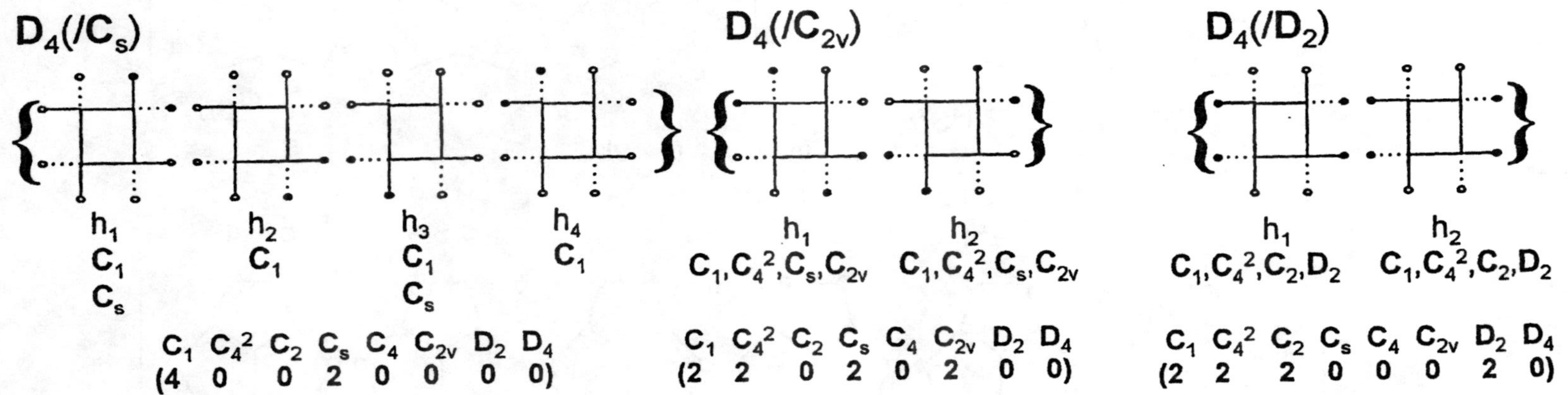

D4(/Cs)
h1
C1
Cs
h2
C1
h3
C1
Cs
h4
C1
C1 C4 2 C2 Cs C4 C2v D2 D4
(4 0 0 2 0 0 0 0)
D4(/C2v)
h1
C1,C4 2,Cs,C2v
h2
C1,C4 2,Cs,C2v
C1 C4 2 C2 Cs C4 C2v D2 D4
(2 2 0 2 0 2 0 0)
D4(/D2)
h1
C1,C4 2,C2,D2
h2
C1,C4 2,C2,D2
C1 C4 2 C2 Cs C4 C2v D2 D4
(2 2 2 0 0 0 2 0)

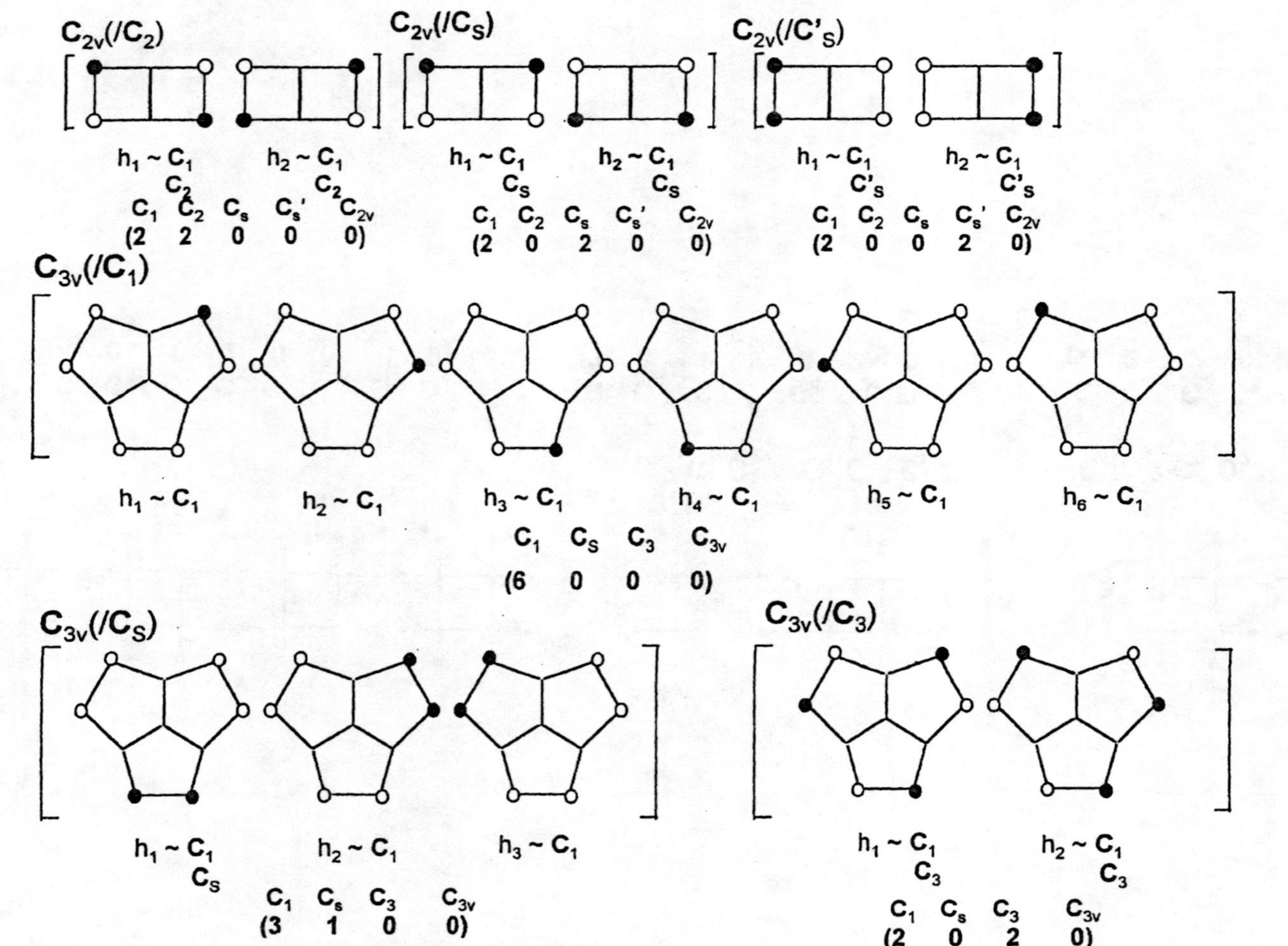

FIG48 Colored graphs (homomers) which model corset representations of C_{2v}, and C_{3v} point-groups. Subgroups which stabilize homomers are indicated together with the resulting mark rows.

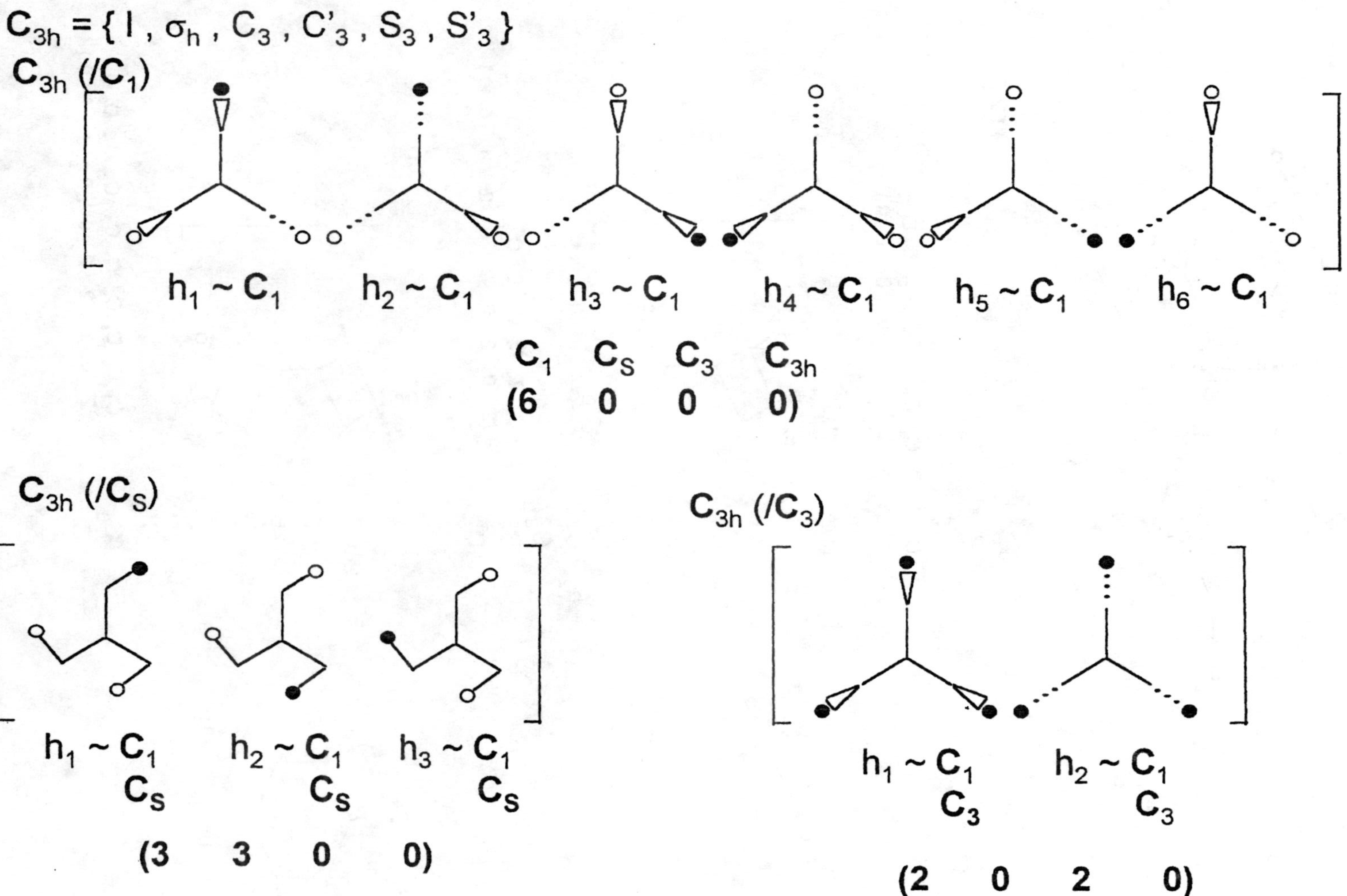

FIG50 Modeling coset representations of C_{3h}. Homomers , their stabilizing subgroups and the relevant mark rows are indicated.

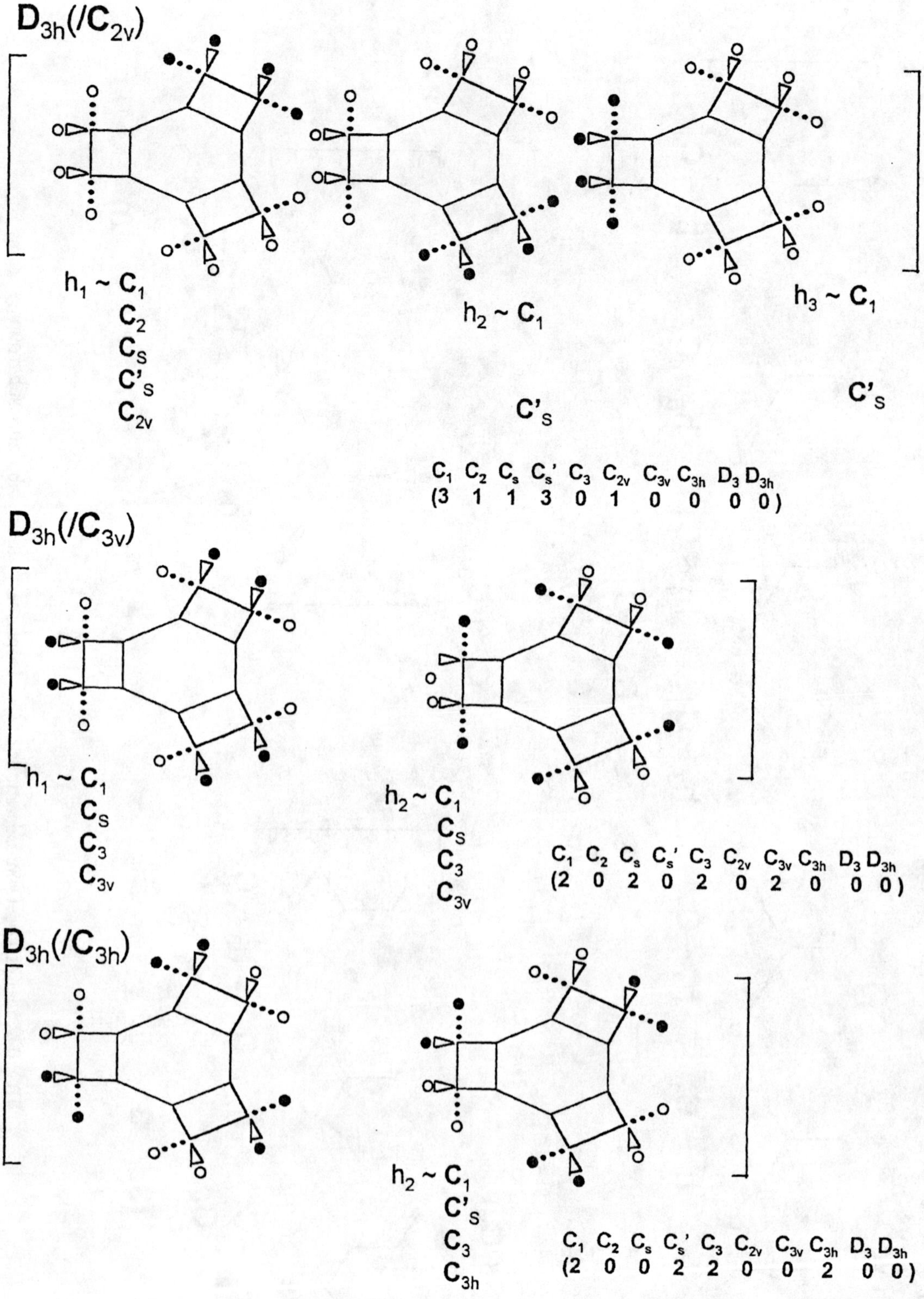

$D_{3h}(/C_{2v})$

$h_1 \sim C_1$
C_2
C_s
C'_s
C_{2v}

$h_2 \sim C_1$
C'_s

$h_3 \sim C_1$
C'_s

C_1	C_2	C_s	C_s'	C_3	C_{2v}	C_{3v}	C_{3h}	D_3	D_{3h}
(3	1	1	3	0	1	0	0	0	0)

$D_{3h}(/C_{3v})$

$h_1 \sim C_1$
C_s
C_3
C_{3v}

$h_2 \sim C_1$
C_s
C_3
C_{3v}

C_1	C_2	C_s	C_s'	C_3	C_{2v}	C_{3v}	C_{3h}	D_3	D_{3h}
(2	0	2	0	2	0	2	0	0	0)

$D_{3h}(/C_{3h})$

$h_2 \sim C_1$
C'_s
C_3
C_{3h}

C_1	C_2	C_s	C_s'	C_3	C_{2v}	C_{3v}	C_{3h}	D_3	D_{3h}
(2	0	0	2	2	0	0	2	0	0)

FIG51 Modeling coset representations of some coset representations of $\mathbf{D_{3h}}$. Homomers , their stabilizing subgroups and the resulting mark rows are indicated.

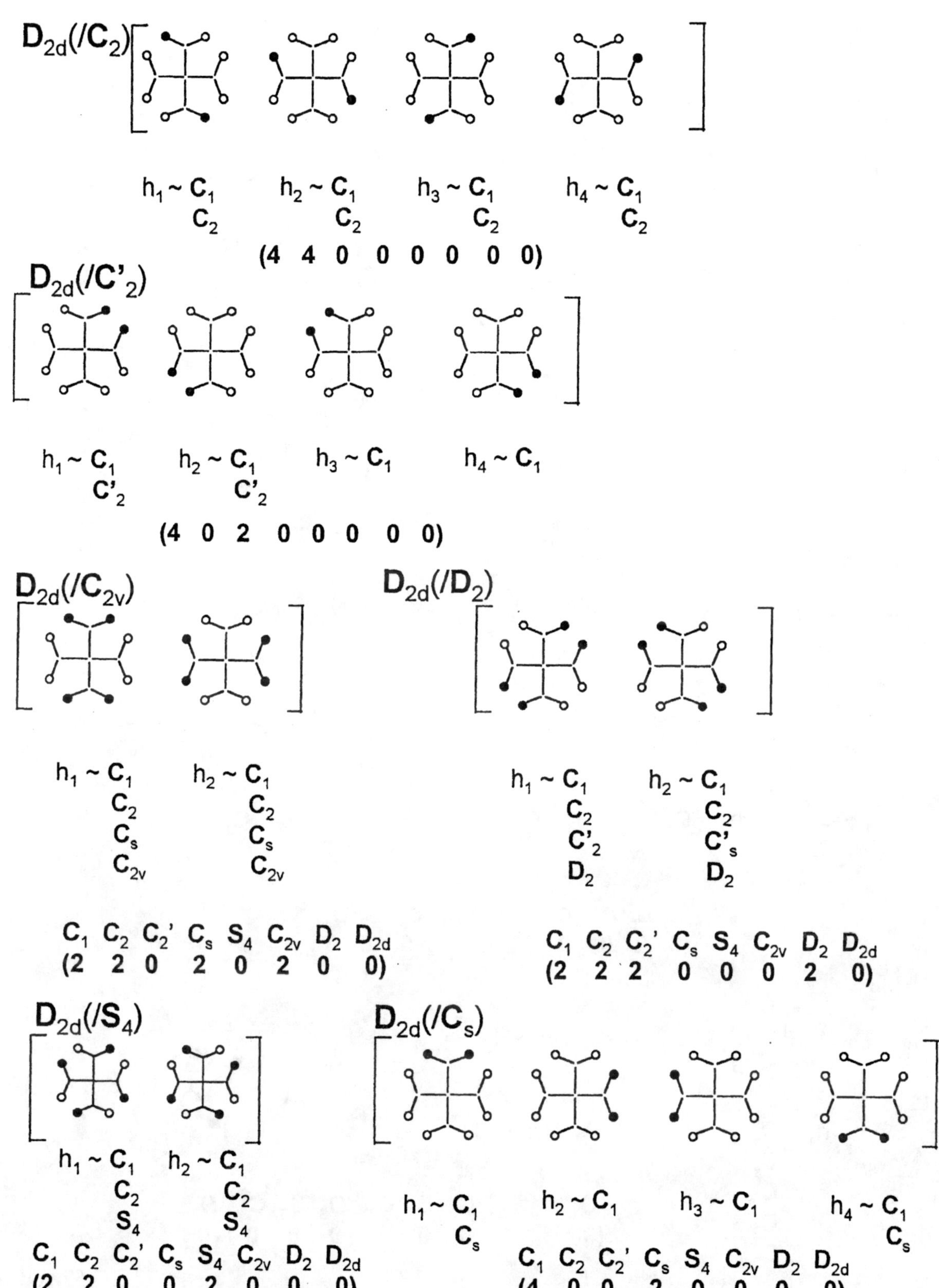

FIG52 Models of selected coset representations of $\mathbf{D_{2d}}$ and $\mathbf{D_{3d}}$, their stabilizing subgroups and mark rows. Homomer models of $\mathbf{D_{2d}}(/C_2)$, $\mathbf{D_{2d}}(/C_{2v})$, $\mathbf{D_{2d}}(/D_2)$, $\mathbf{D_{3d}}(/C_2)$ and $\mathbf{D_{3d}}(/C_3)$, are stabilized under the same set of subgroups in each case.

D3d(/C2)
h1 ~ C1 Cs
h2 ~ C1 Cs
h3 ~ C1 Cs
h4 ~ C1 Cs
h5 ~ C1 Cs
h6 ~ C1 Cs
(6 0 0 6 0 0 0 0 0 0)
D3d(/C3)
h1 ~ C1 C3
h2 ~ C1 C3
h3 ~ C1 C3
h4 ~ C1 C3
C1 Cs Cs' C2 C3 C2h D3 S6 C3v D3d
(4 0 0 0 4 0 0 0 0 0)
D3d(/Cs)
h1 ~ C1 Cs
h2 ~ C1
h3 ~ C1
h4 ~ C1 Cs
h5 ~ C1
h6 ~ C1
C1 Cs Cs' C2 C3 C2h D3 C3v S3 D3d
(6 2 0 0 0 0 0 0 0 0)
D3d(/C2h)
h1 ~ C1 Cs C's C2 C2v
h2 ~ C1 C2
h3 ~ C1 C2
C1 Cs Cs' C2 C3 C2h D3 S6 C3v D3d
(3 1 1 3 0 1 0 0 0 0)

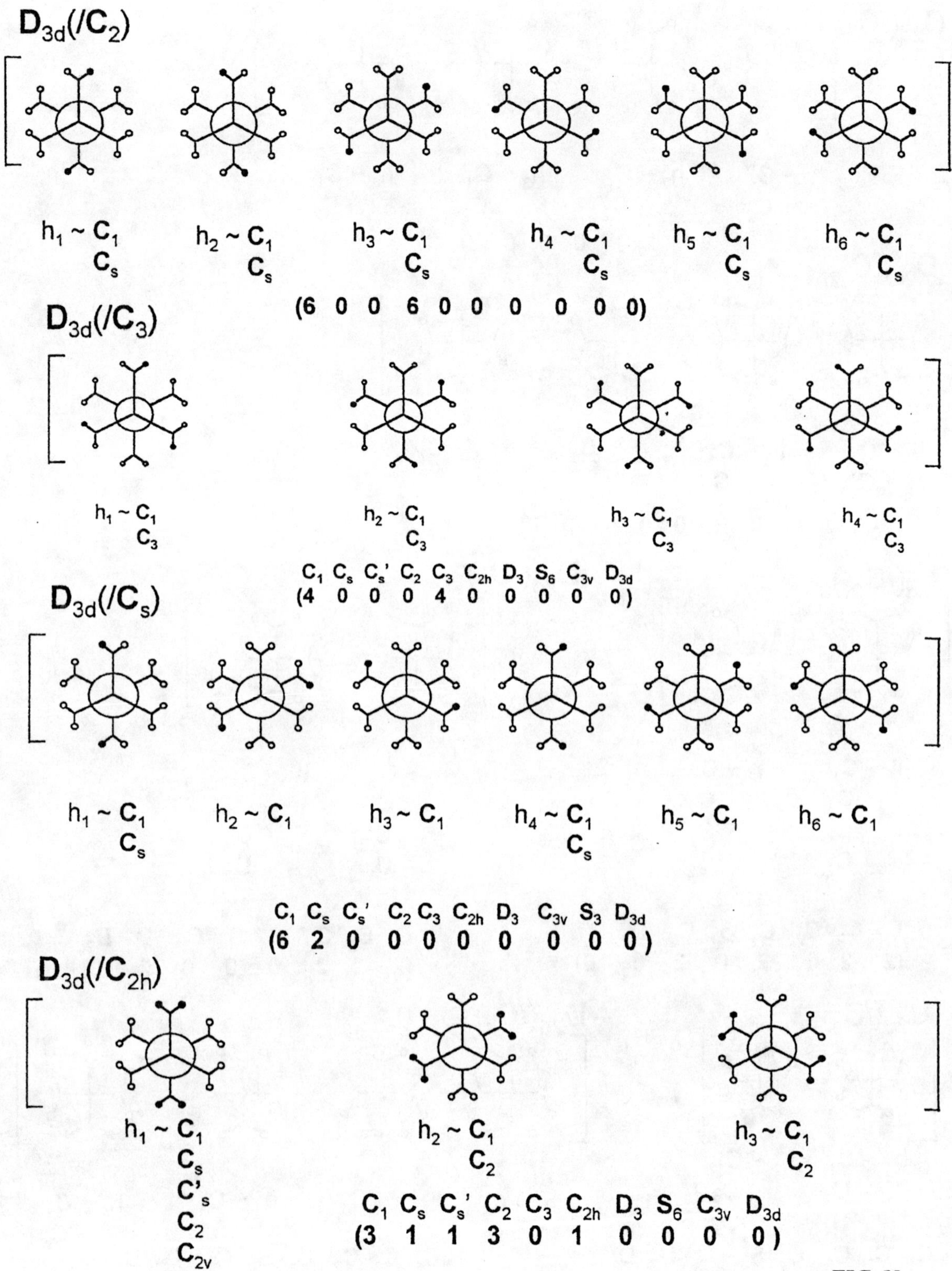

FIG 52 cont

$S_4 = \{ I, S_4, C_2, S_4^3 \}$

$S_4(/C_1)$

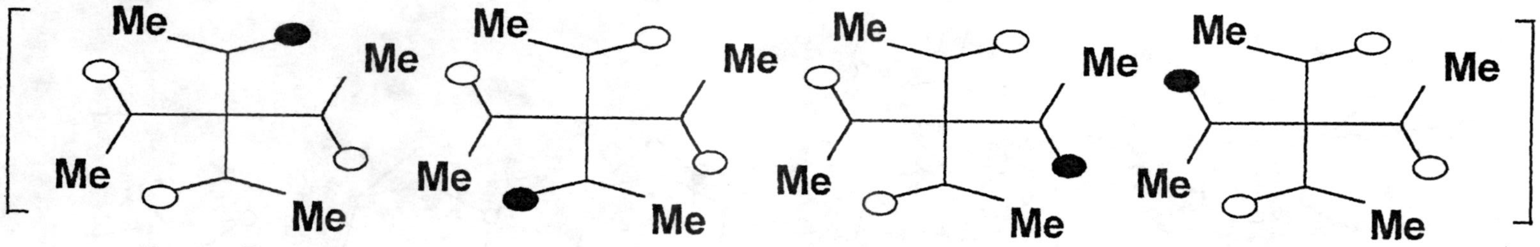

$h_1 \sim C_1$ $h_2 \sim C_1$ $h_3 \sim C_1$ $h_4 \sim C_1$

C_1	C_S	S_4
(4	0	0)

$S_4(/C_2)$

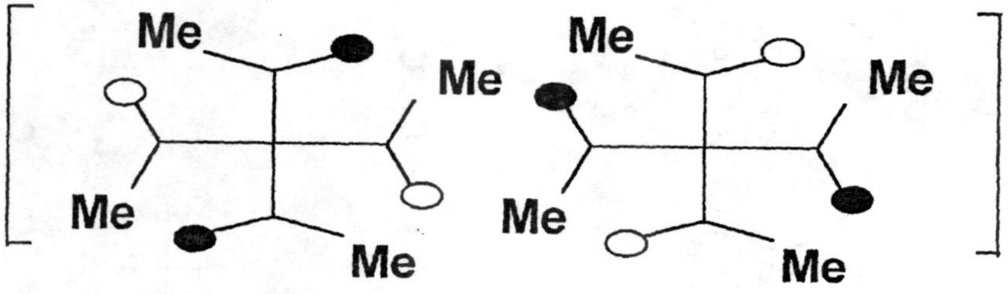

$h_1 \sim C_1$, C_2 $h_2 \sim C_1$, C_2

(2 2 0)

FIG53 Homomers of coset representations of S_4, their stabilizing subgroups and mark rows.

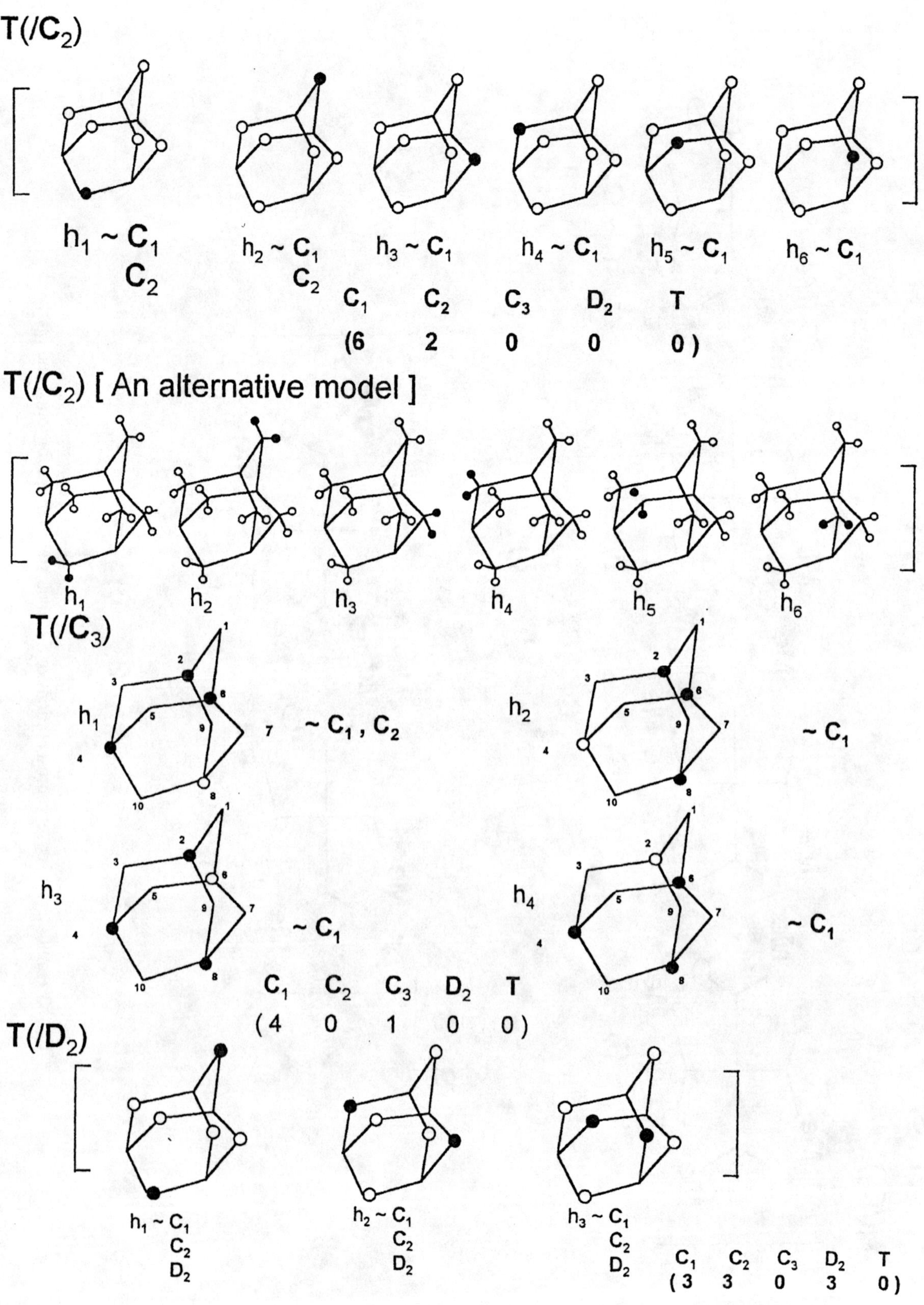

FIG54 Homomers of some coset representations of **T**, their stabilizing subgroups and mark rows. The homomers of **T(/D$_2$)** are stabilized by the same set of subgroups.

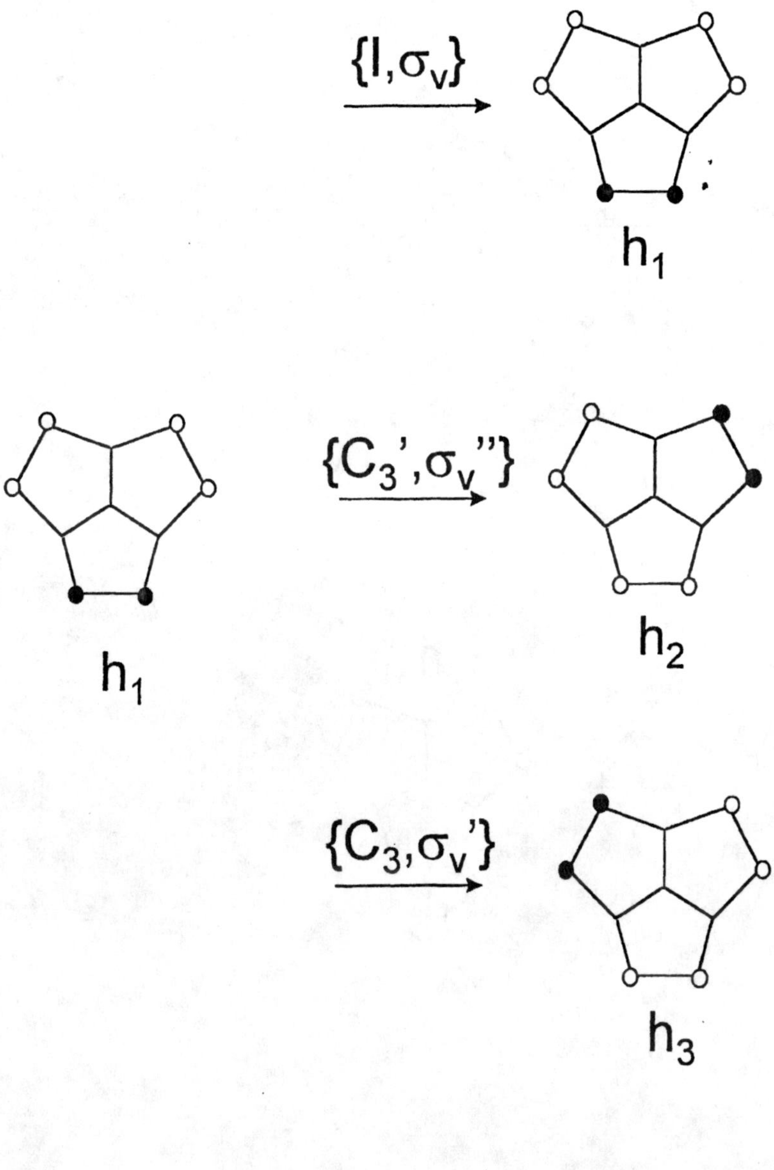

FIG55 Generation of the set of homomers of $C_{3v}(/C_s)$.

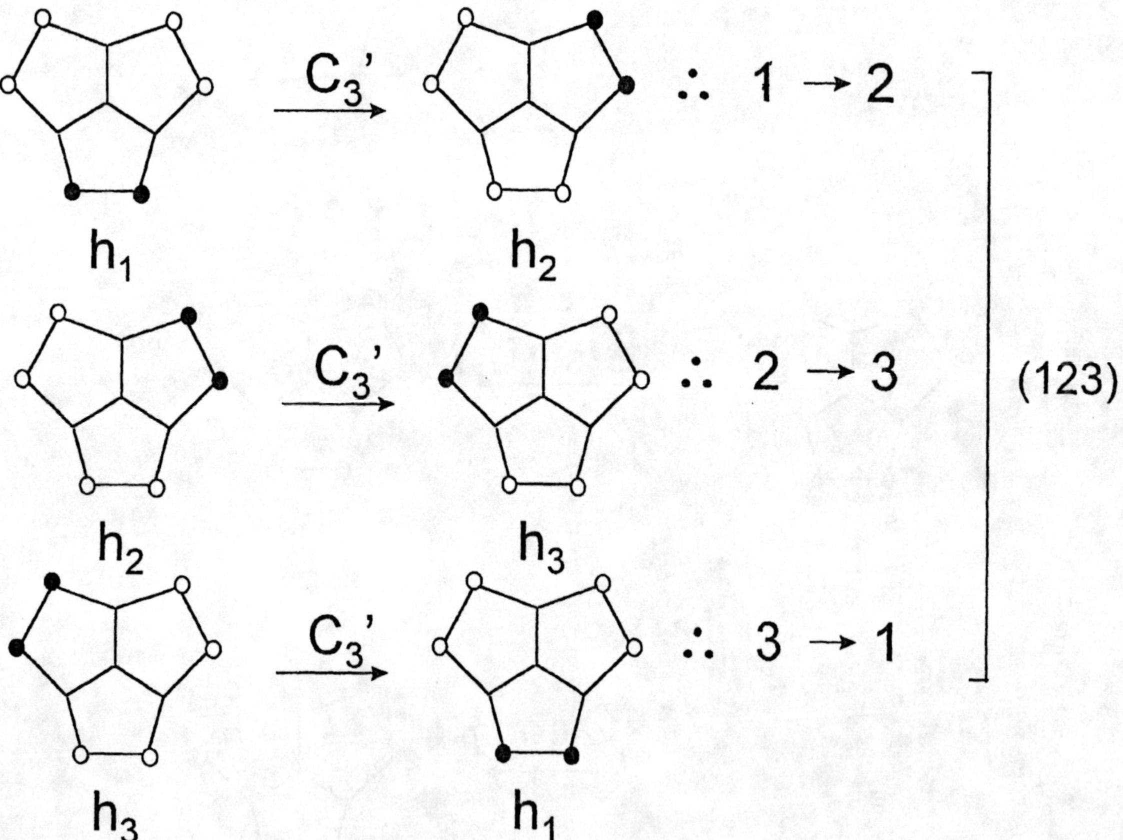

FIG56 Permutation induced when C_3' is applied to the three homomers which model $C_{3v}(/C_s)$; c.f. eqn. (44).

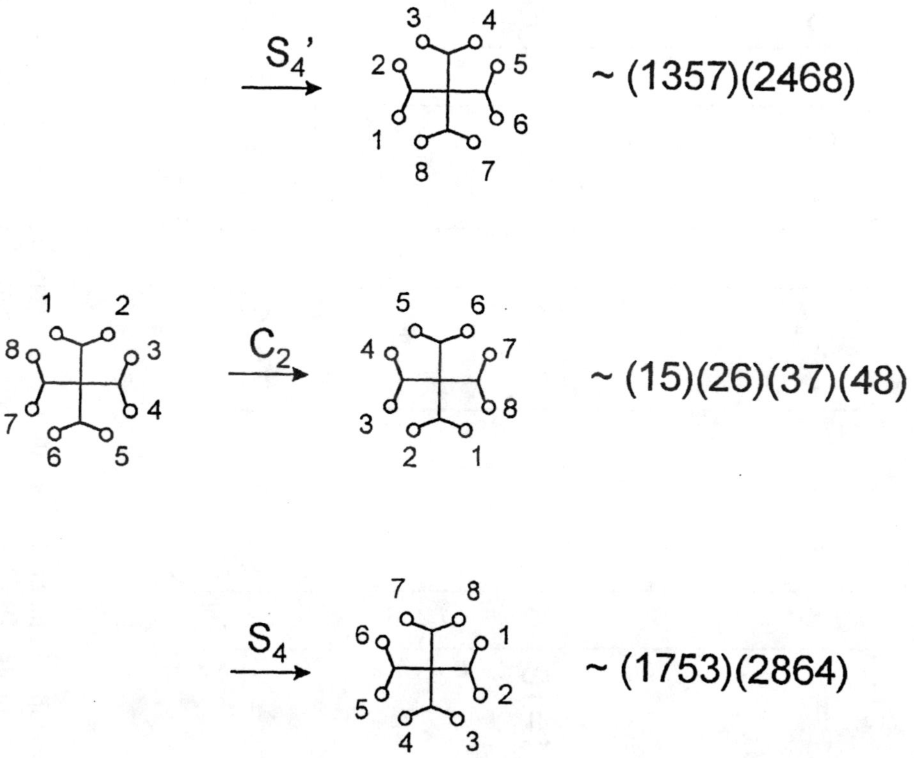

$\xrightarrow{S_4'}$ ~ (1357)(2468)

$\xrightarrow{C_2}$ ~ (15)(26)(37)(48)

$\xrightarrow{S_4}$ ~ (1753)(2864)

FIG57 The effects of the elements of symmetry of S_4 on the identity representation $D_{2d}(/D_{2d})$. The resulting permutations lead to the "color" identities given by eqn. (218).

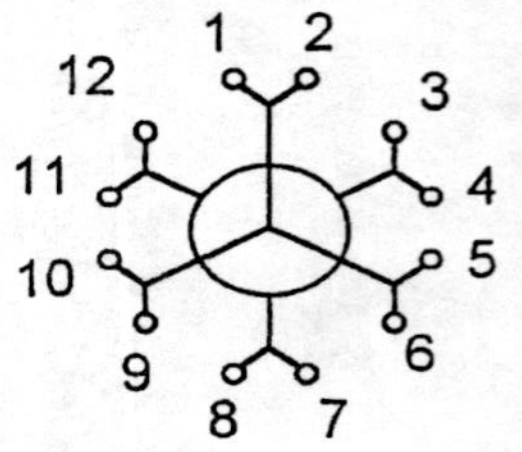

$$\xrightarrow{C_2}\begin{pmatrix}1=9\\ \\ 1=10\\ \\ 1=2\end{pmatrix};\begin{pmatrix}2=10\\ \\ 2=9\\ \\ 9=10\end{pmatrix};\begin{pmatrix}7=5\\ \\ 7=4\\ \\ 7=12\end{pmatrix};\begin{pmatrix}6=8\\ \\ 3=8\\ \\ 8=11\end{pmatrix};\begin{pmatrix}3=11\\ \\ 6=11\\ \\ 3=6\end{pmatrix};\begin{pmatrix}4=12\\ \\ 5=12\\ \\ 4=5\end{pmatrix}$$

Rows generated by C_2 (top), σ_d (middle), σ_1 (bottom).

FIG58 Generation of the "color" equalities of eqn.(218) which result in the homomer which models $\mathbf{D_{3d}}(/\mathbf{C_{2h}})$. This homomer is labeled h_1 under $\mathbf{D_{3d}}(/\mathbf{C_{2h}})$ in **FIG**58.

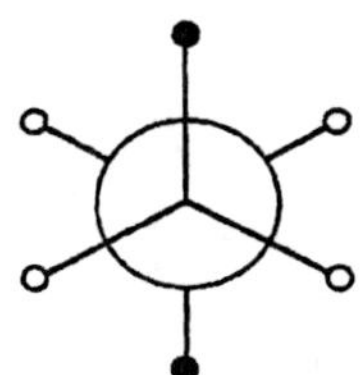

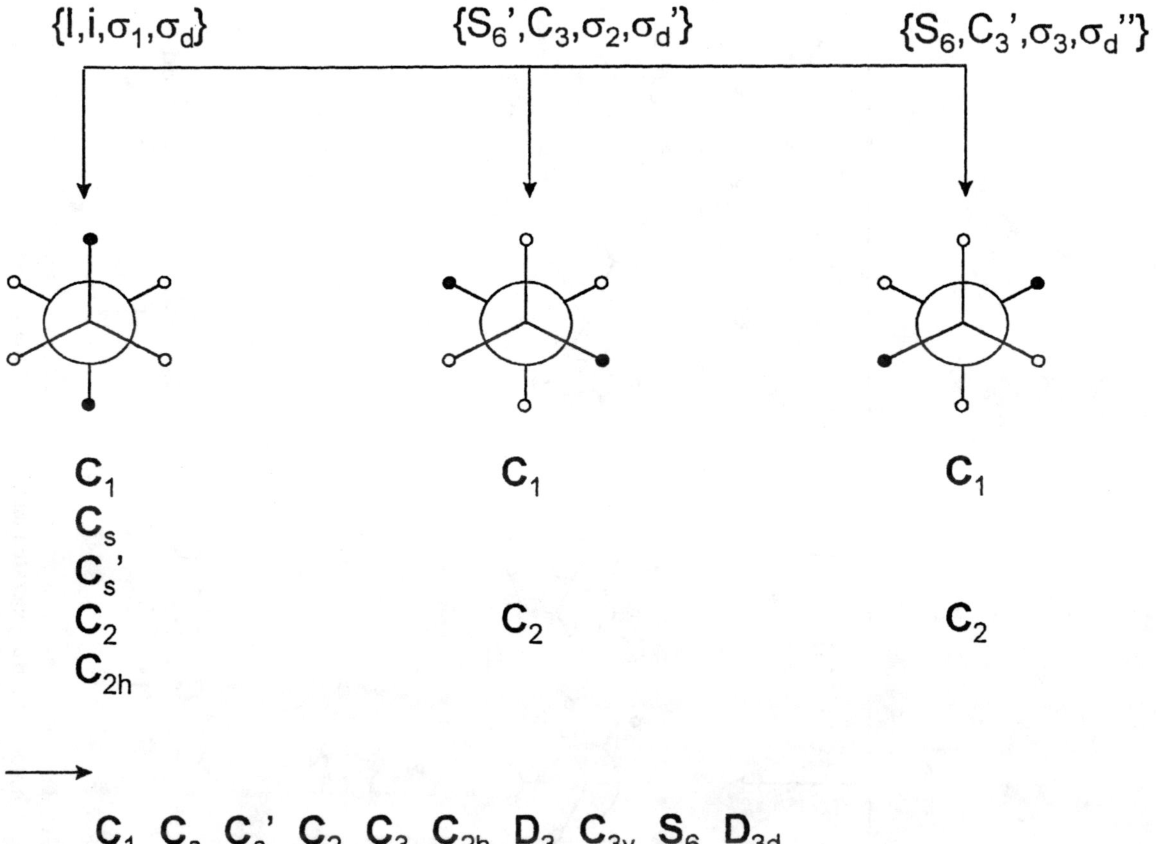

C_1	C_s	C_s'	C_2	C_3	C_{2h}	D_3	C_{3v}	S_6	D_{3d}
(3	1	1	3	0	1	0	0	0	0)

FIG59 A wrong model for $\mathbf{D_{3d}}(/\mathbf{C_2})$ leading to wrong row of marks. Observe that this model remains invariant also under $\mathbf{C_s}$ and $\mathbf{C_s}$'. The correct model is shown in **FIG**60.

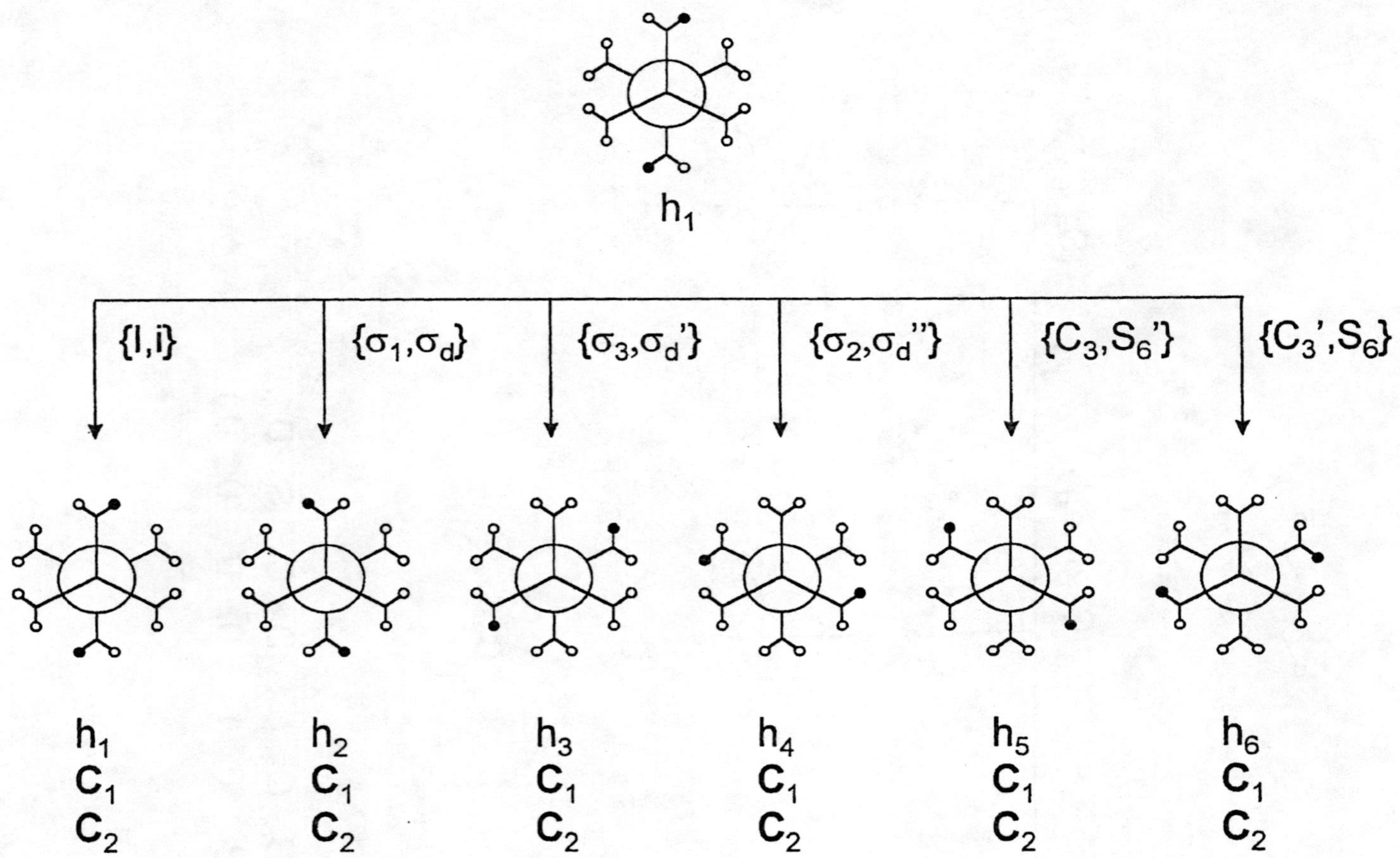

h_1

$\{I, i\}$ $\{\sigma_1, \sigma_d\}$ $\{\sigma_3, \sigma_d'\}$ $\{\sigma_2, \sigma_d''\}$ $\{C_3, S_6'\}$ $\{C_3', S_6\}$

h_1	h_2	h_3	h_4	h_5	h_6
C_1	C_1	C_1	C_1	C_1	C_1
C_2	C_2	C_2	C_2	C_2	C_2

C_1	C_s	C_s'	C_2	C_3	C_{2h}	D_3	C_{3v}	S_6	D_{3d}
(6	0	0	6	0	0	0	0	0	0)

FIG60 The correct modeling of $\mathbf{D_{3d}}(/C_2)$. The figure is intended to show how other homomers are generated and the row of marks modeled.

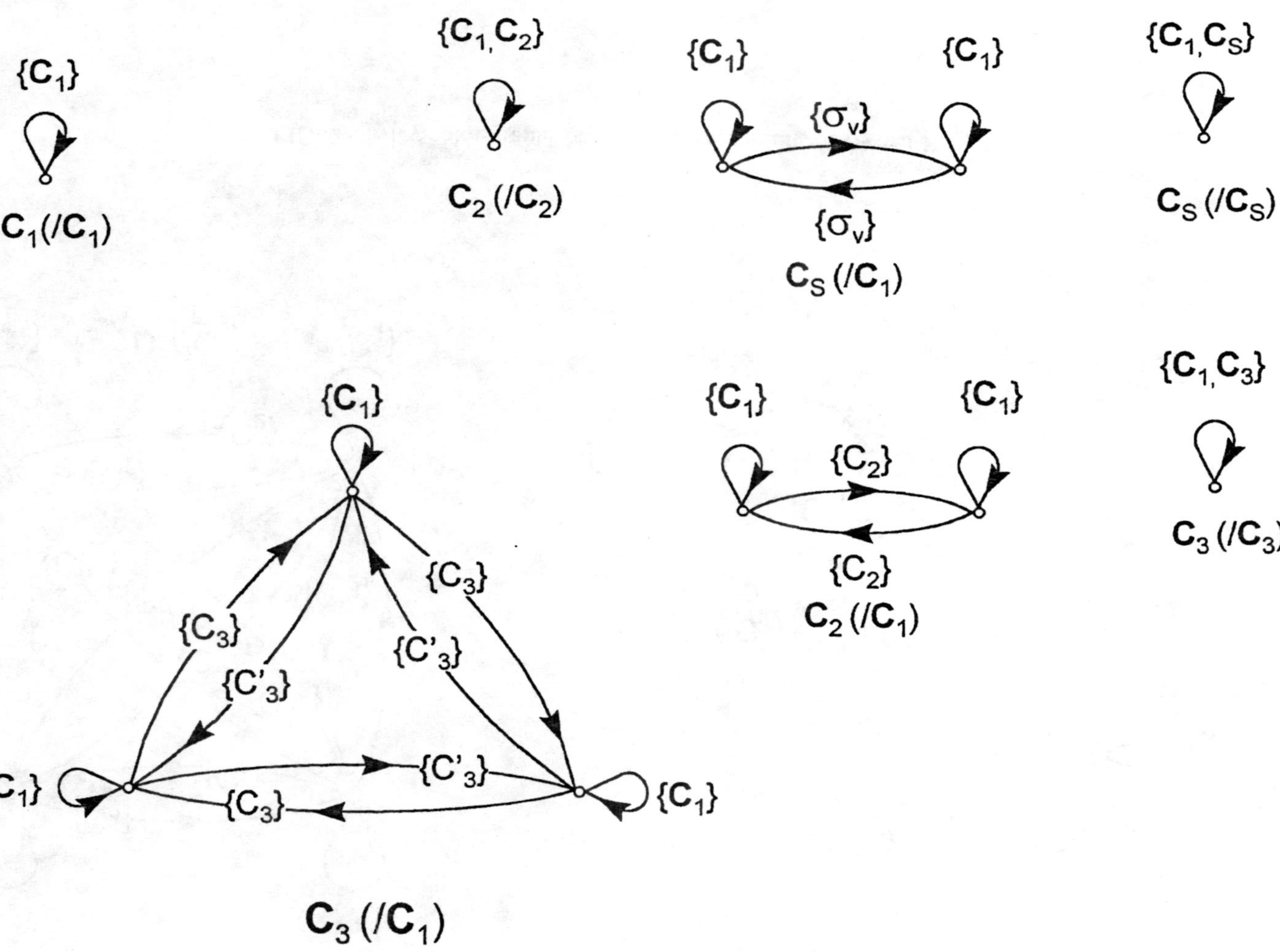

FIG61 Cayley color graphs of coset representations of C_1 , C_2 , C_s and C_3 groups.

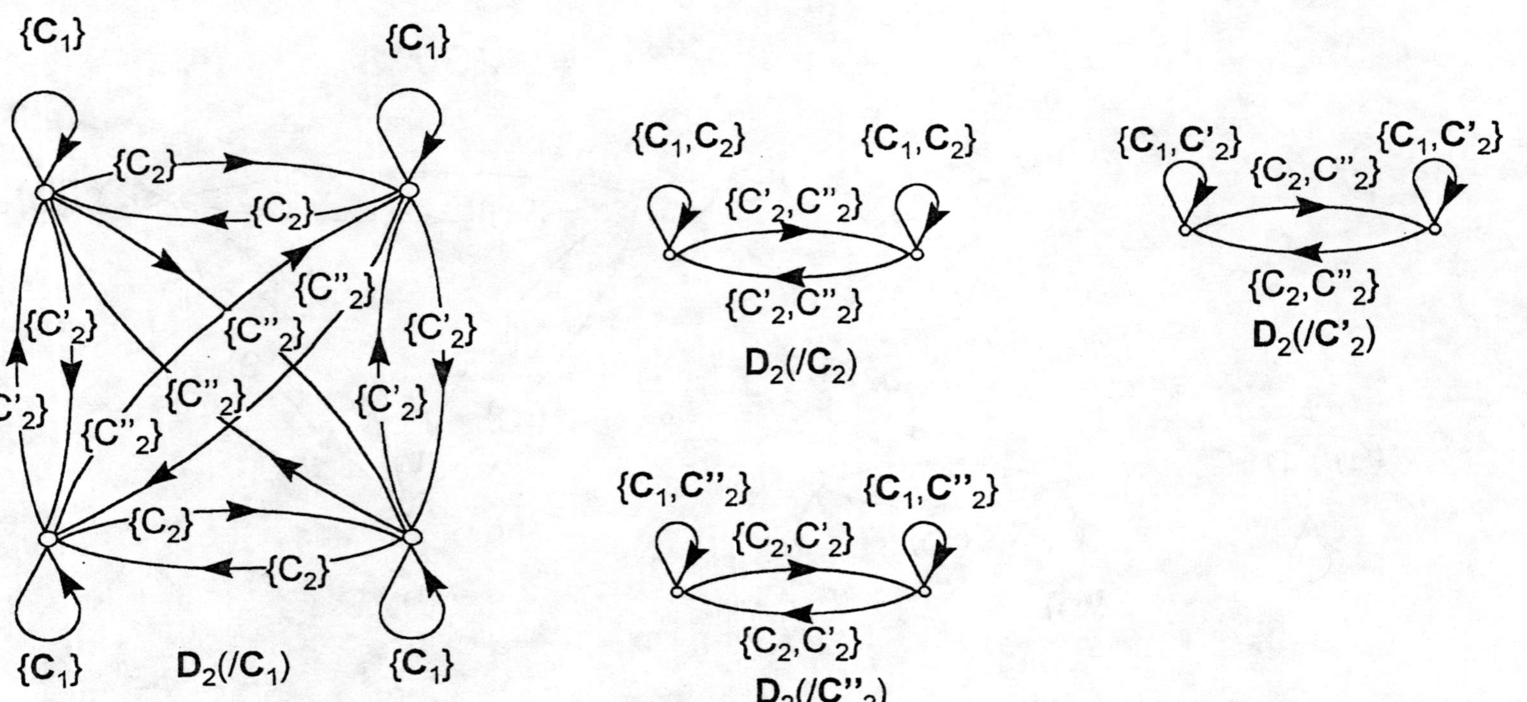

FIG62 Cayley color graphs of coset representations of $\mathbf{D}_2$ and $\mathbf{D}_3$ groups.

{C1,C2}
{C'2,C3}
{C''2,C3}
{C''2,C'3}
{C'2,C'3}
{C1,C2}
{C2,C'3}
{C2,C3}
{C1,C2}
D3(/C2)
{C1,C3}
{C1,C3}
{C2,C'2,C''2}
{C2,C'2,C''2}
D3(/C3)

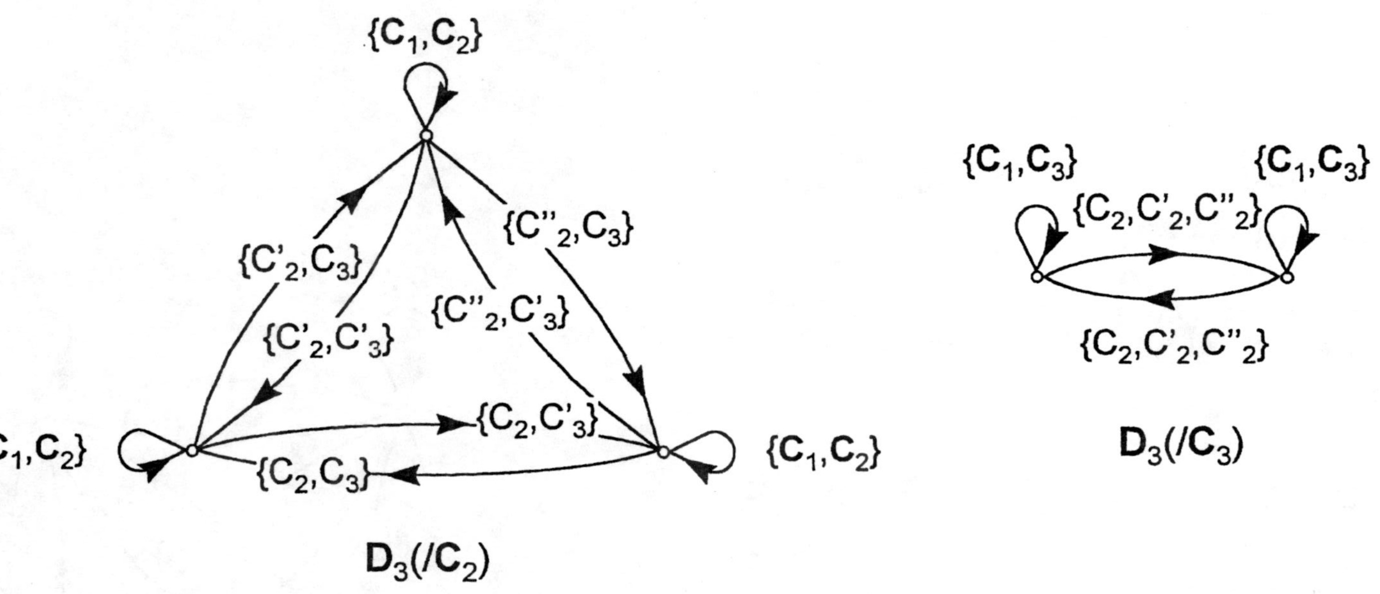

FIG 62 cont.

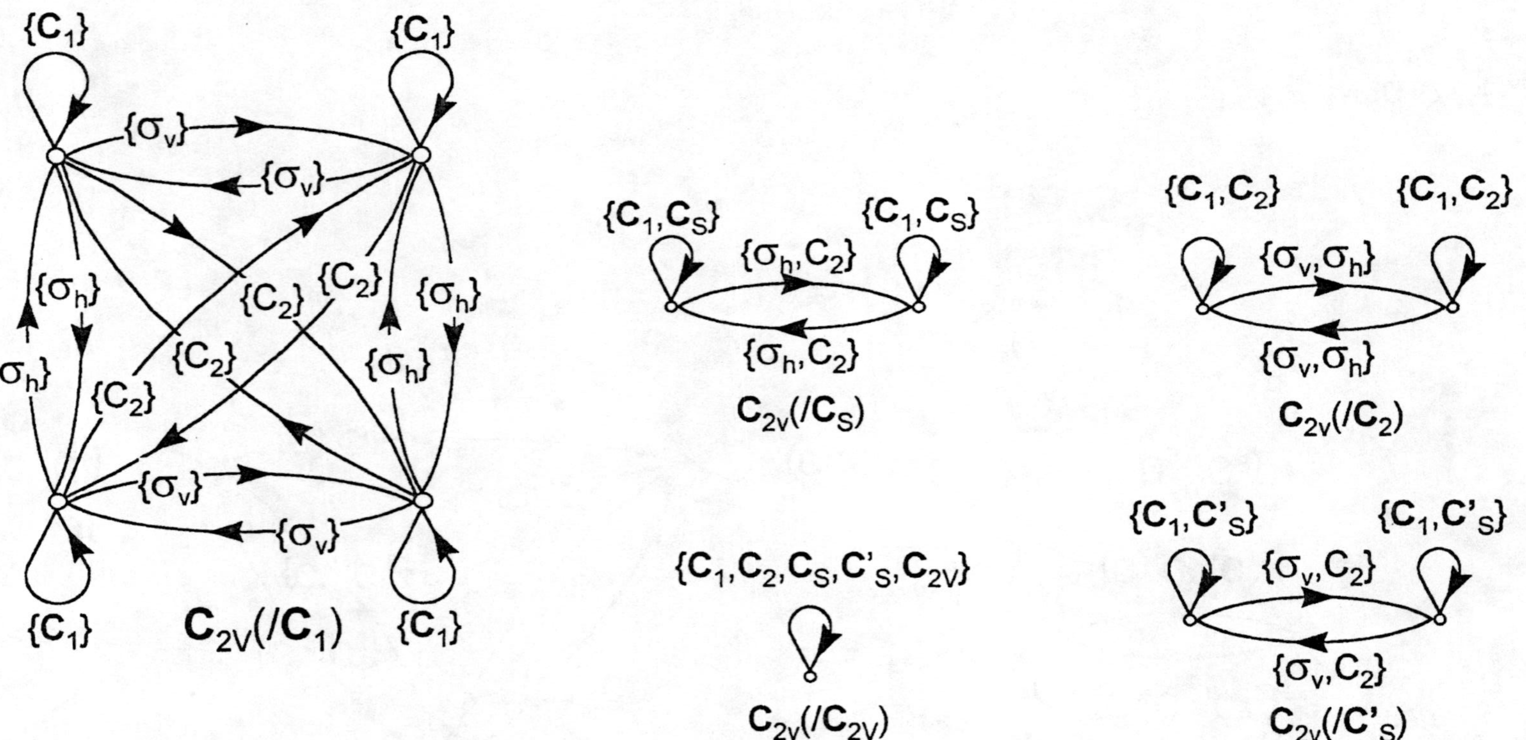

FIG63 Cayley color graphs of coset representations of C_{2v} and C_{3v} point-groups.

{C1,CS}
{C'3,σ'v}
{C'3,σ''v}
{σ'v,C'3}
{C'3,σ''v}
{C1,C''S}
{C'3,σv}
{C'3,σv}
{C1,C'S}
C3v(/CS)
{C1,C3}
{C1,C3}
{σv,σ'v,σ''v}
{σv,σ'v,σ''v}
C3v(/C3)

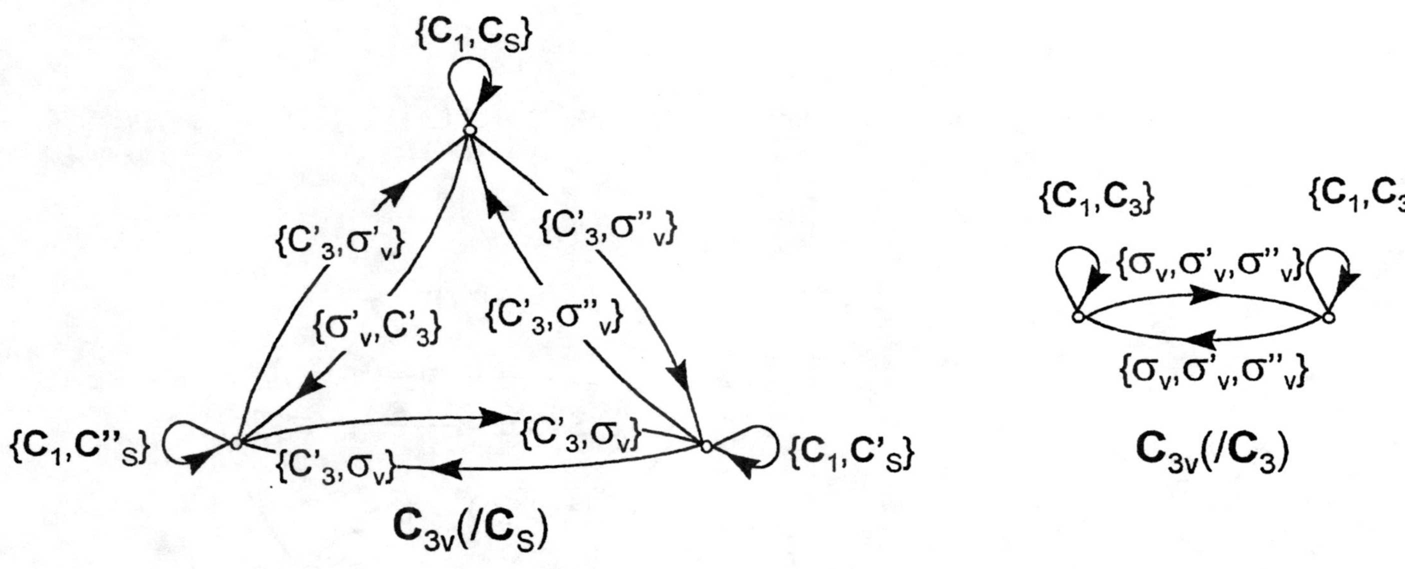

FIG 63 cont.

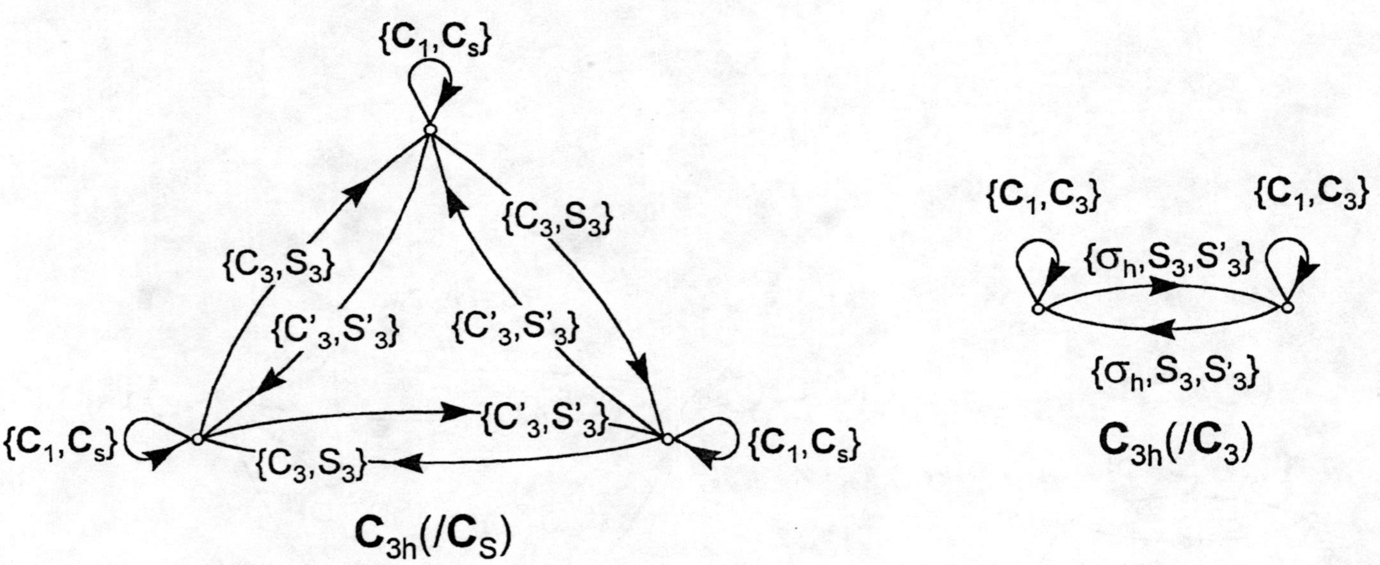

FIG64 Cayley graphs of two coset representations of C_{3h}.

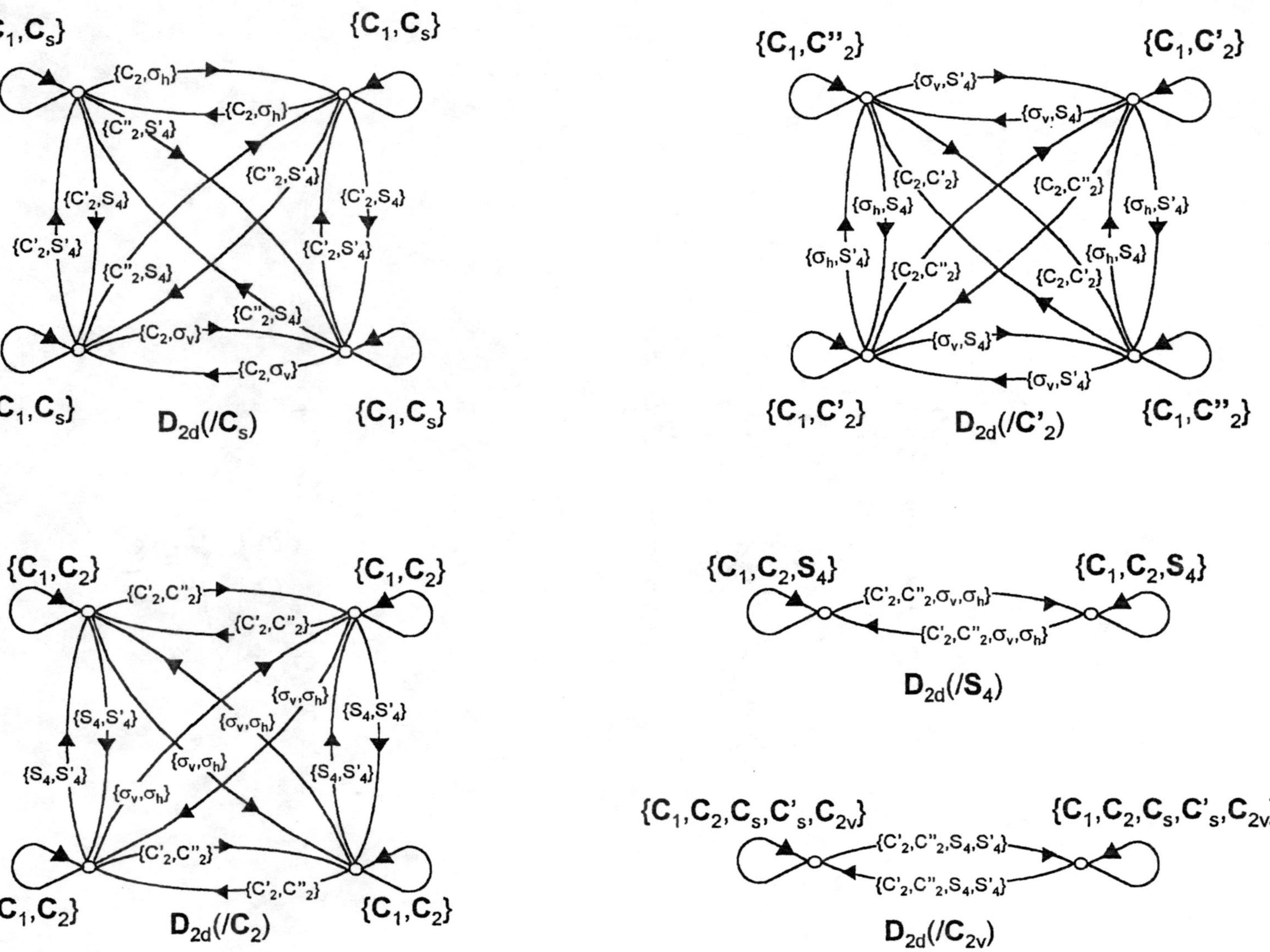

FIG65 Cayley graphs of some coset representations of $\mathbf{D_{2d}}$.

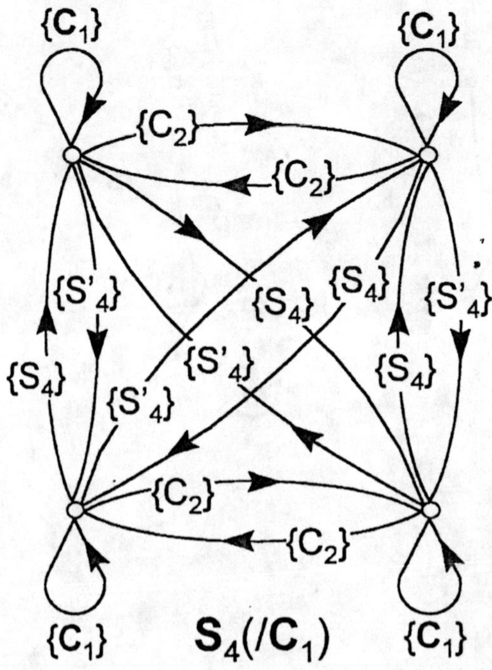

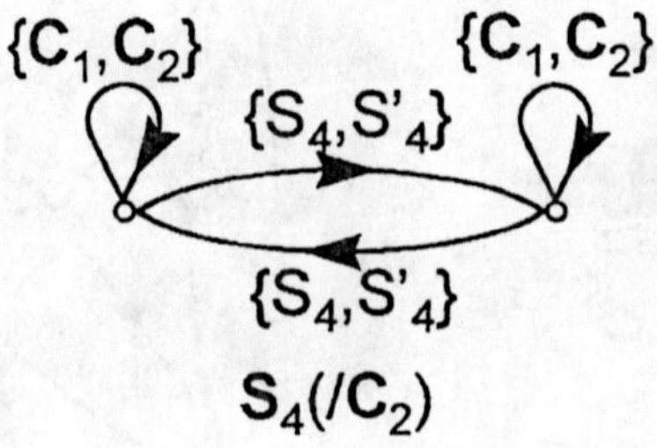

FIG66 Cayley graphs of coset representations of $\mathbf{S}_4$.

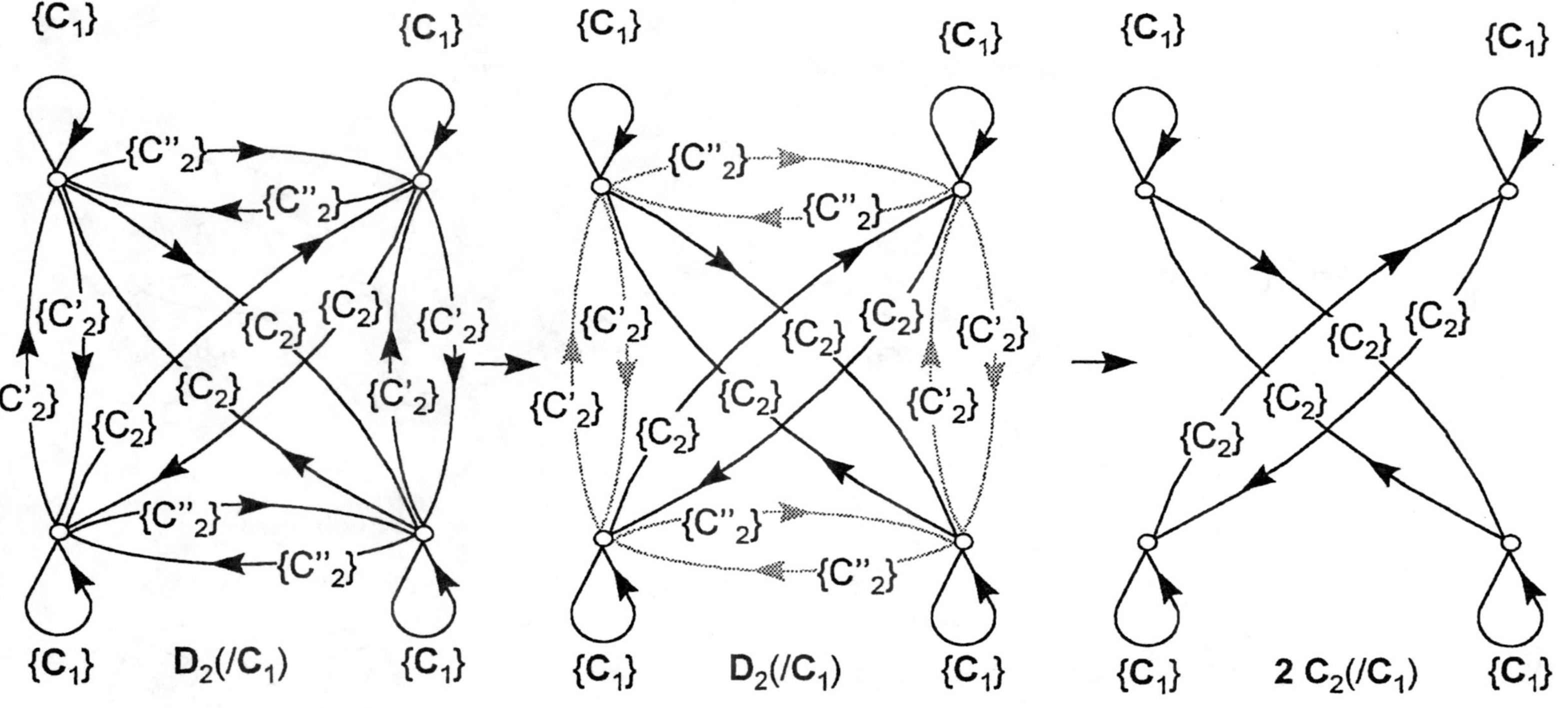

$D_2(/C_1) \downarrow C_2 = 2\, C_2(/C_1)$

FIG67 Illustration of modeling of the subduction equation $\mathbf{D_{2d}}(/C_1) \downarrow C_2 = 2\, C_2(/C_1)$ using the pruning method , c.f. section 3.2.1 In the intermediate step , the arcs to be pruned out are printed in gray. The resulting two copies of the Cayley graph of $C_2(/C_1)$ are to be compared with **FIG**61.

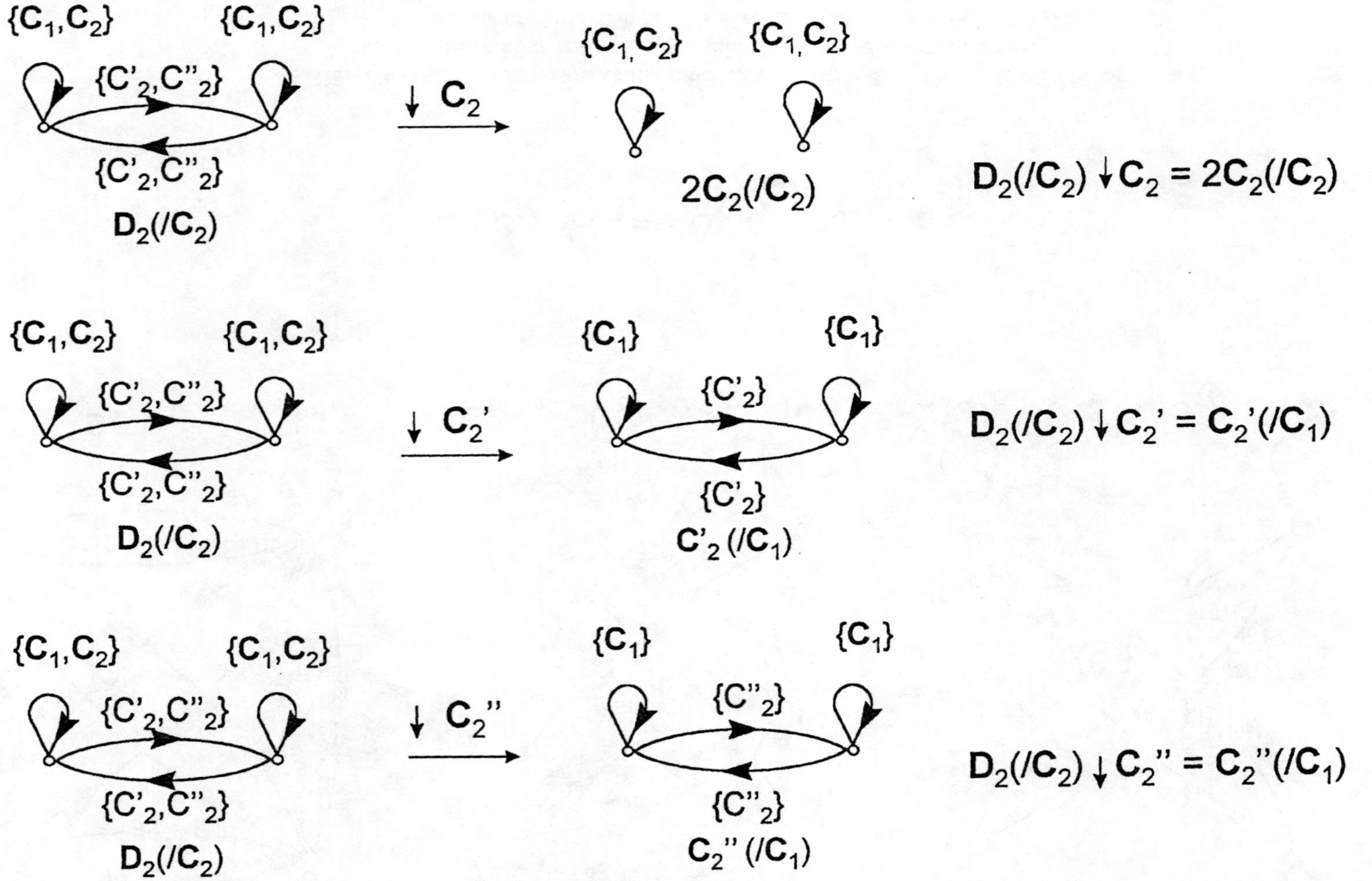

FIG68 Modeling subduction of $\mathbf{D}_2(/\mathbf{C}_2)$ by several subgroups of $\mathbf{D}_2$. (c.f. **FIG**62).

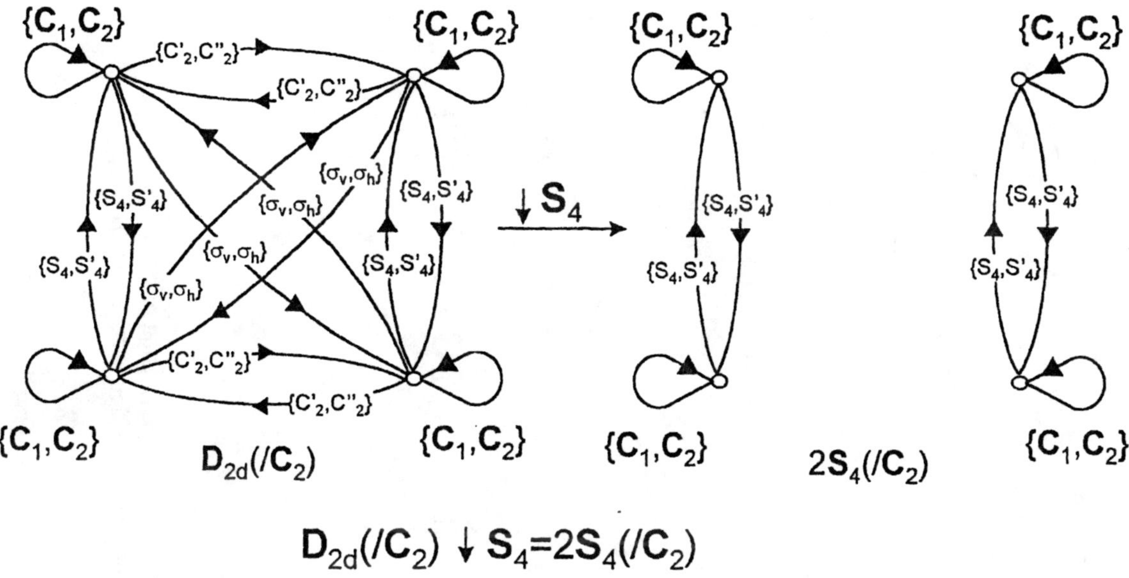

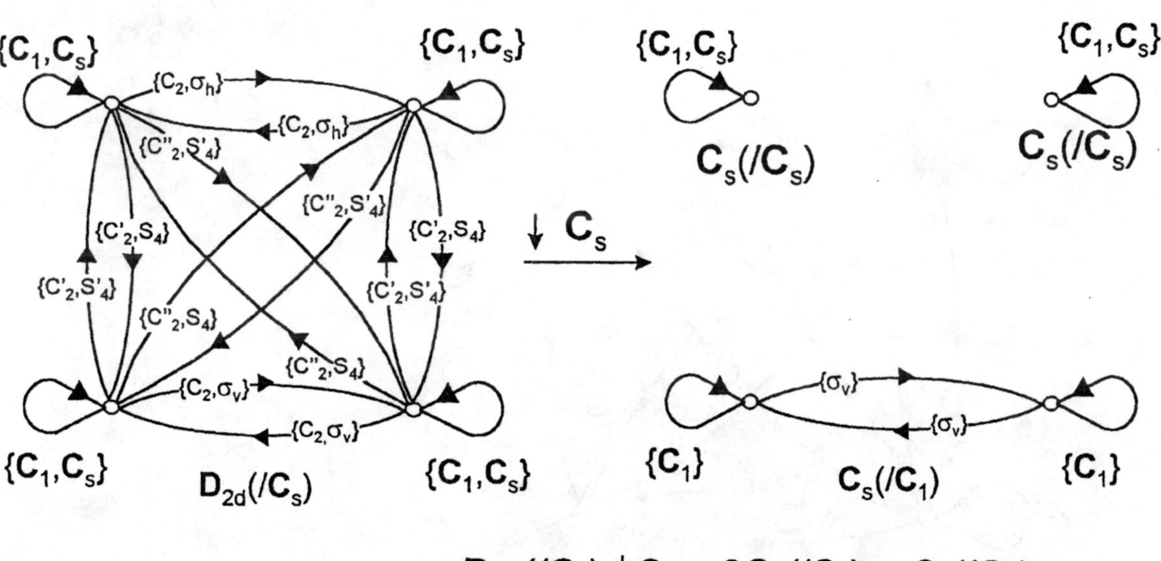

FIG69 The pruning method applied to Cayley graphs of $\mathbf{D_{2d}}$(/C_2) and $\mathbf{D_{2d}}$(/C_s) which model subductions by $\mathbf{S_4}$ and $\mathbf{C_s}$ subgroups respectively. [c.f. section 3.2.1]. The resulting pruned out graphs are to be identified from **FIGs.** 61 and 66.

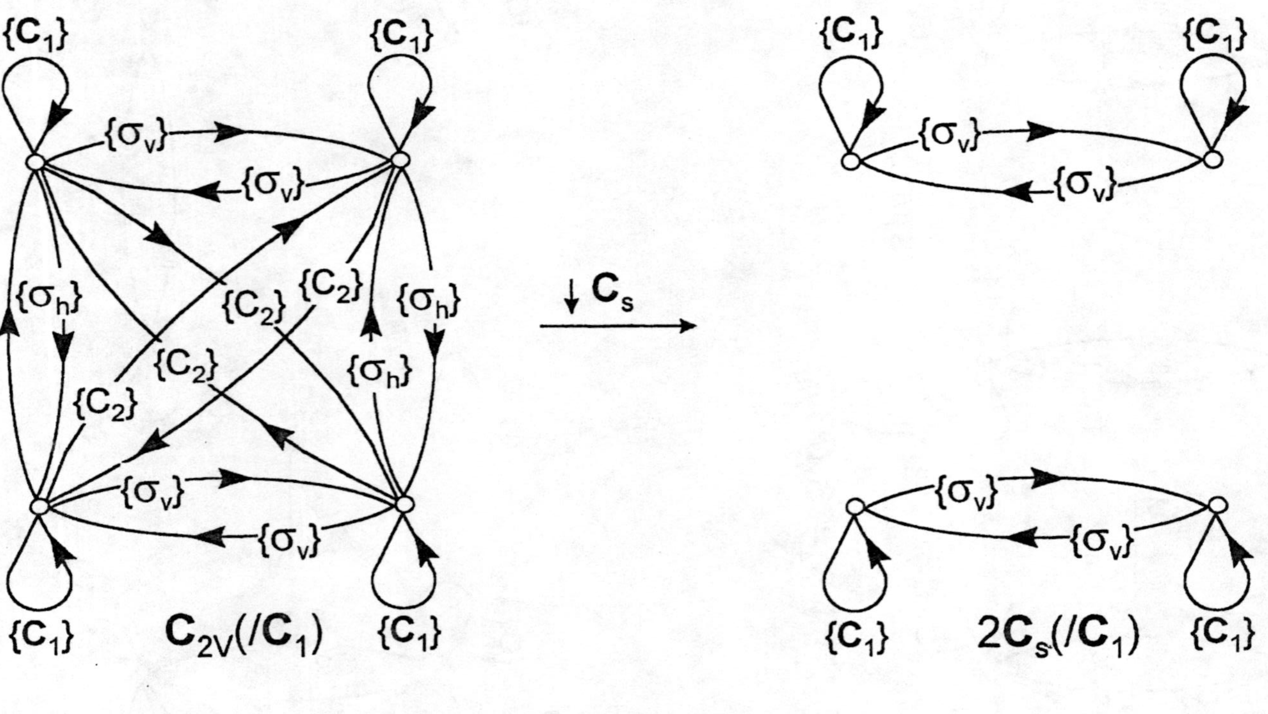

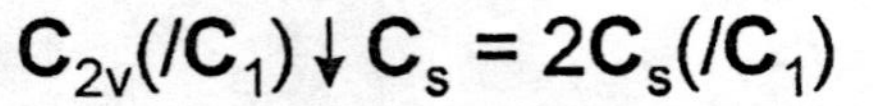

$$C_{2v}(/C_1) \downarrow C_s = 2C_s(/C_1)$$

FIG70 Modeling the subduction eqn. $C_{2v}(/C_1) \downarrow C_s = 2C_s(/C_1)$ using pruning method (section 3.2.1). The resulting pruned-out graphs are identified in **FIG**61.

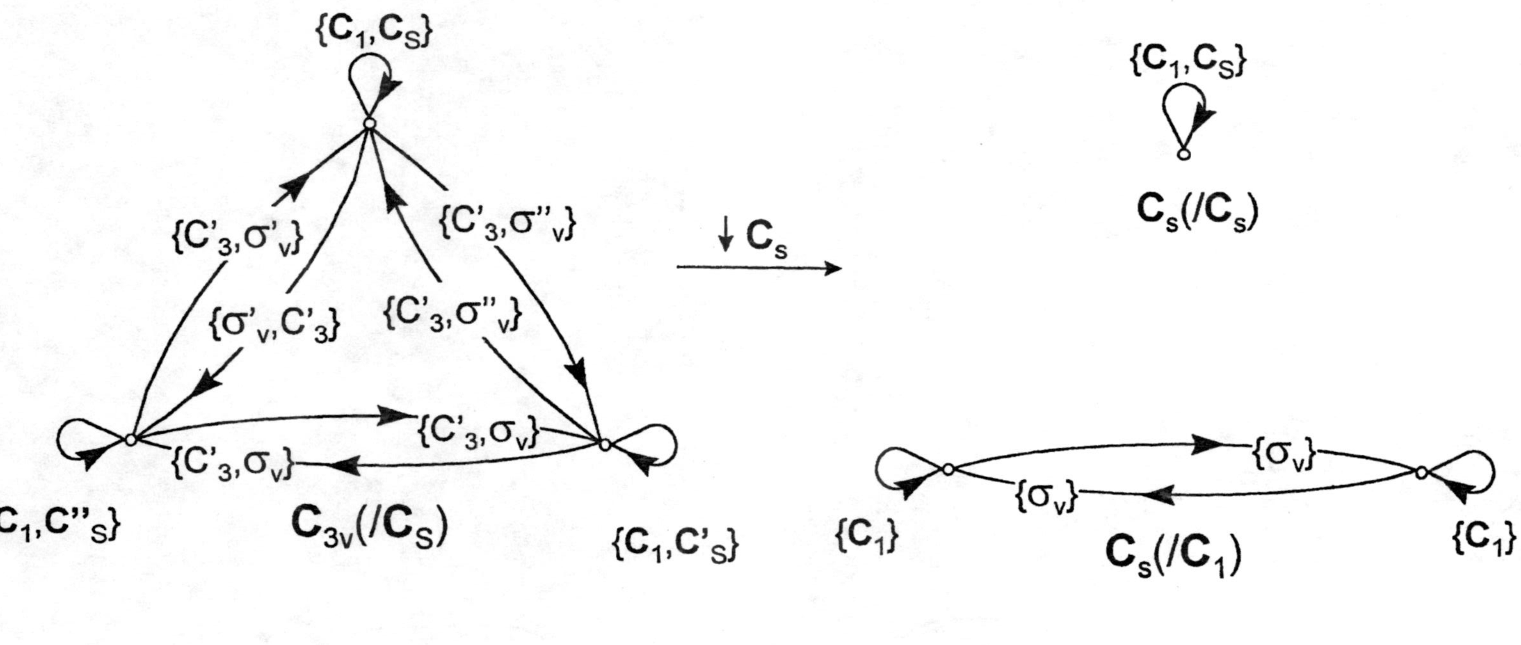

$$C_{3v}(/C_s)\downarrow C_s = C_s(/C_s) + C_s(/C_1)$$

FIG71 Modeling subduction of $C_{3v}(/C_s)$ by C_s using pruning technique (section 3.2.1). The resulting auxiliary graphs are identified in **FIG**61.

$$\begin{pmatrix} \{C_1\} & \{C_2\} \\ \{C_2\} & \{C_1\} \end{pmatrix}$$

M$[C_2\ (/C_1)]$
(2 0)

$[\{C_1\ C_2\}]$
M $[C_2\ (/C_2)]$
(1 1)

$$\begin{pmatrix} \{C_1\} & \{\sigma_v\} \\ \{\sigma_v\} & \{C_1\} \end{pmatrix}$$

M $[C_s\ (/C_1)]$
(2 0)

$[\{C_1, C_s\}]$
M $[C_s\ (/C_s)]$
(1 1)

$$\begin{pmatrix} \{C_1\} & \{C_2\} & \{S_4\} & \{S_4'\} \\ \{C_2\} & \{C_1\} & \{S_4'\} & \{S_4\} \\ \{S_4'\} & \{S_4\} & \{C_1\} & \{C_2\} \\ \{S_4\} & \{S_4'\} & \{C_2\} & \{C_1\} \end{pmatrix}$$

M $[S_4\ (/C_1)]$
(4 0 0)

$$\begin{pmatrix} \{C_1, C_2\} & \{S_4, S_4'\} \\ \{S_4, S_4'\} & \{C_1, C_2\} \end{pmatrix}$$

M $[S_4\ (/C_1)]$
(2 2 0)

$(\{C_1, C_2, S_4\})$
M $[S_4\ (/S_4)]$
(1 1 1)

$$\begin{pmatrix} \{C_1\} & \{C_2\} & \{\sigma_v\} & \{\sigma_h\} \\ \{C_2\} & \{C_1\} & \{\sigma_h\} & \{\sigma_v\} \\ \{\sigma_v\} & \{\sigma_h\} & \{C_1\} & \{C_2\} \\ \{\sigma_h\} & \{\sigma_v\} & \{C_2\} & \{C_1\} \end{pmatrix}$$

M $[C_{2v}\ (/C_1)]$
(4 0 0 0 0)

$$\begin{pmatrix} \{C_1, C_2\} & \{\sigma_v, \sigma_h\} \\ \{\sigma_v, \sigma_h\} & \{C_1, C_2\} \end{pmatrix}$$

M $[C_{2v}\ (/C_2)]$
(2 2 0 0 0)

FIG72 Matrix representations of the coset representations of some of the basic point groups. Mark rows which are indicated can be found by listing the number of subgroups which occur along the main diagonal of the relevant **M**.

$$\begin{pmatrix} \{C_1,C_s\} & \{\sigma_h,C_2\} \\ \{\sigma_h,C_2\} & \{C_1,C_s\} \end{pmatrix}$$

M [C_{2v} (/C_s)]

(2 0 2 0 0)

$$\begin{pmatrix} \{C_1,C_s'\} & \{\sigma_v,C_2\} \\ \{\sigma_v,C_2\} & \{C_1,C_s'\} \end{pmatrix}$$

M [C_{2v} (/C_s')]

(2 0 0 2 0)

$$\begin{pmatrix} \{C_1\} & \{C_2\} & \{C_2'\} & \{C_2''\} \\ \{C_2\} & \{C_1\} & \{C_2''\} & \{C_2'\} \\ \{C_2'\} & \{C_2''\} & \{C_1\} & \{C_2\} \\ \{C_2''\} & \{C_2'\} & \{C_2\} & \{C_1\} \end{pmatrix}$$

M [C_{2v} (/C_s')]

(4 0 0 0 0)

$$\begin{pmatrix} \{C_1,C_2\} & \{C_2',C_2''\} \\ \{C_2',C_2''\} & \{C_1,C_2\} \end{pmatrix}$$

M [D_2 (/C_2)]

(2 2 0 0 0)

$$\begin{pmatrix} \{C_1,C_2'\} & \{C_2,C_2''\} \\ \{C_2,C_2''\} & \{C_1,C_2'\} \end{pmatrix}$$

M [D_2 (/C_2'')]

(2 0 0 2 0)

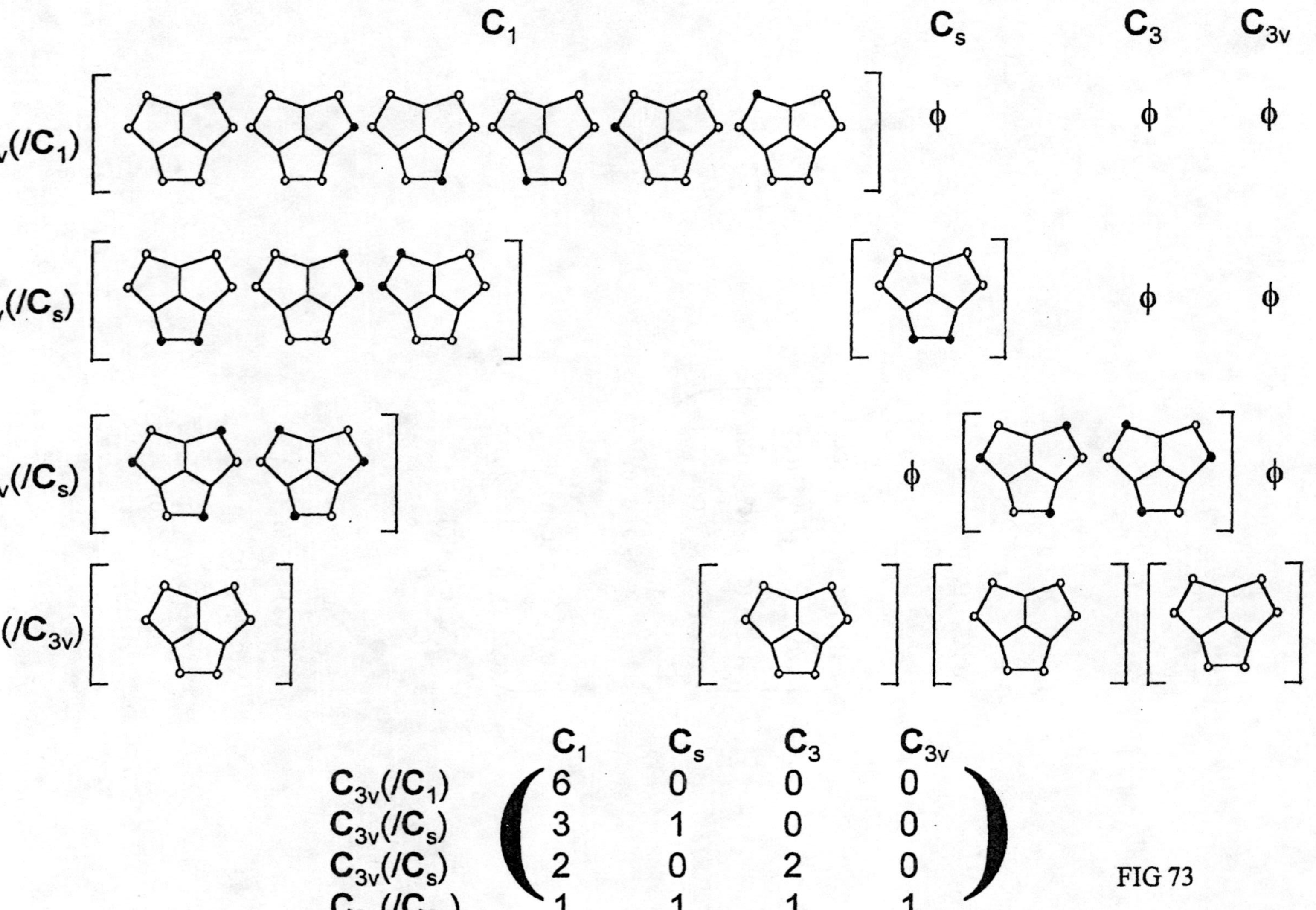

	C_1	C_s	C_3	C_{3v}
$C_{3v}(/C_1)$	6	0	0	0
$C_{3v}(/C_s)$	3	1	0	0
$C_{3v}(/C_s)$	2	0	2	0
$C_{3v}(/C_{3v})$	1	1	1	1

FIG 73

FIG73 A graphical model for the mark table of C_{3v} along with the table given in its standard conventions. The model makes the meaning of a mark much more transparent. For example the 6 in the uppermost left corner indicates 6 colored graphs which are fixed under C_1. These graphs are drawn and so on for other numbers in the table.

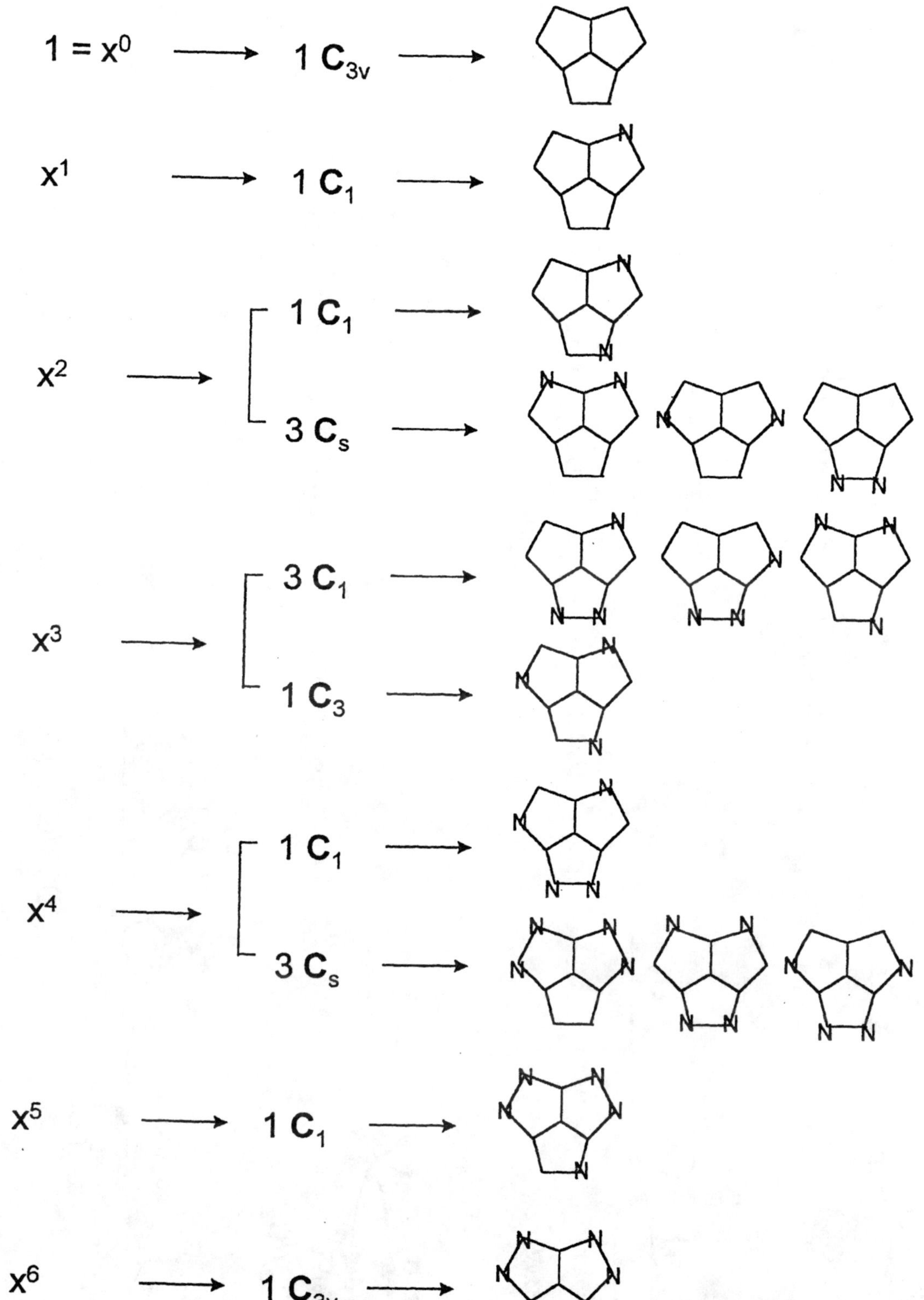

$1 = x^0 \longrightarrow 1\ C_{3v} \longrightarrow$

$x^1 \longrightarrow 1\ C_1 \longrightarrow$

$x^2 \longrightarrow$ [$1\ C_1 \longrightarrow$; $3\ C_s \longrightarrow$]

$x^3 \longrightarrow$ [$3\ C_1 \longrightarrow$; $1\ C_3 \longrightarrow$]

$x^4 \longrightarrow$ [$1\ C_1 \longrightarrow$; $3\ C_s \longrightarrow$]

$x^5 \longrightarrow 1\ C_1 \longrightarrow$

$x^6 \longrightarrow 1\ C_{3v} \longrightarrow$

FIG74 The derivatives given by the isomer-count matrix , eqn.(255) where the number of N atoms = the power of X.

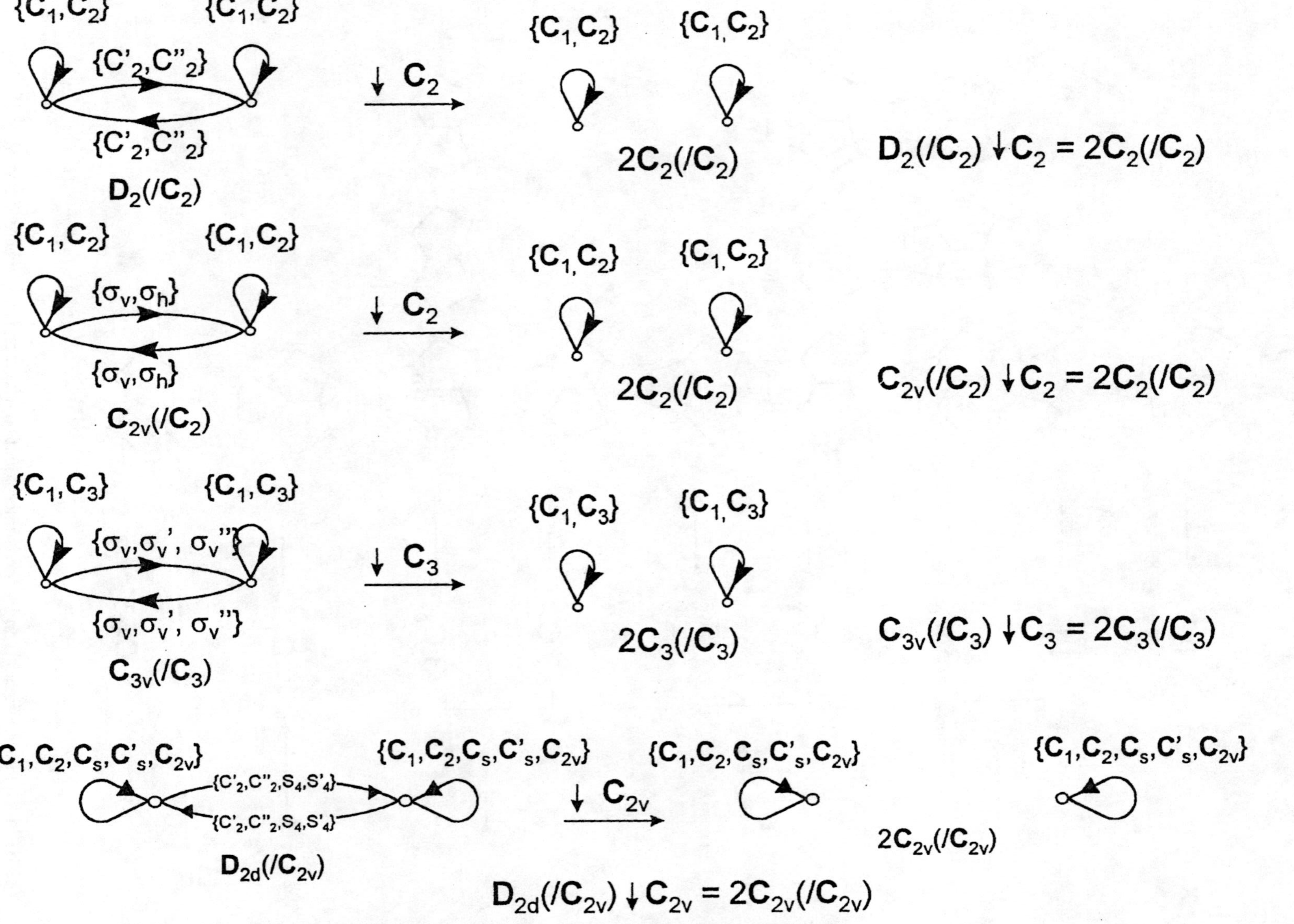

FIG75 Exemplification of eqn.(267) where all homomers of a given coset representation remain invariant under the same set of subgroups. [c.f. **FIGs.** 48 ,52].

Index